Lecture Notes in Chemistry

Edited by G. Berthier, M. J. S. Dewar, H. Fischer
K. Fukui, H. Hartmann, H. H. Jaffé, J. Jortner
W. Kutzelnigg, K. Ruedenberg, E. Scrocco, W. Zeil

5

Ramon Carbo
Josep M. Riera

A General SCF Theory

Springer-Verlag
Berlin Heidelberg New York 1978

Authors
Ramon Carbo
Division of Theoretical Chemistry
Department of Chemistry
University of Alberta
Edmonton, Alberta, Canada T6G 2G2
and
Departamento de Biomatematicas
Facultad de Medicina
Universidad Autonoma de Barcelona
Bellaterra/Spain

Josep M. Riera
Centro de Calculo and
Seccion de Quimica Cuantica
Departamento de Quimica Organica
Instituto Quimico de Sarria
Barcelona-17/Spain

Library of Congress Cataloging in Publication Data

Carbó, Ramón.
A general SCF theory.

(Lecture notes in chemistry ; 5).
Bibliography: p.
Includes index.
1. Self consistent field theory. I. Riera, Josep M., joint author. II. Title.
QD461.C26 541'.24 77-19004

ISBN-13: 978-3-540-08535-5 e-ISBN-13: 978-3-642-93075-1
DOI: 10.1007/978-3-642-93075-1

Softcover reprint of the hardcover 1st edition 1978

2152/3140-543210

Assassins de raons i de vides
que mai no tingueu repòs en
cap dels vostres dies i que
en la mort us persegueixin
les nostres memòries

Campanades a morts
Lluis Llac (1976)

Tots sabíem on era la Paret,
però ignoràvem què hi havia
darrera ...

Tocant a mà.
J.V. Foix (1972)

Introduction

We live in a molecular world, almost closed shell in nature, and for this reason Chemistry has been a science dealing with closed shell molecules.
However, the high degree of experimental sophistication reached in the past decade has made more apparent the role of open shell structures in chemical research.
A parallel phenomenon can be observed in the development of SCF theory, where closed shell molecular calculations at any level of complexity compose the main body of references which can be obtained in Quantum Chemistry today.
Besides the linkage between experimental and theoretical behaviour, there are, obviously, other reasons which can be attached to a lack of molecular open shell calculations. Among others, there was no connection between closed or open shell theoretical treatments. In this manner, many computational features used by closed shell connoisseurs have not been extended to other computational areas. Since the work of Roothaan in 1960, the open shell molecular landscape has been, theoretically, a very closed one. Further development of SCF theory, which has led to an outburst of multiconfigurational procedures, has paid no, or very faint, attention to the interconnection between these SCF theory advanced features, the open shell framework and closed shell common practice. A good theoretical goal, generally speaking, and in particular inside SCF theory, may consist of a procedure which can be used to solve a given chemical problem, within the physical and approximate limits of the theory.
A restricted SCF formalism has been chosen here. From this point of view the present schemes should be seen as one of the last steps in Roothaan's formalism. As a consequence, all theoretical developments are mainly based on projection operator algebra, which is used in order to obtain a unique coupling operator and a unique SCF equation. This choice is not arbitrary, but follows from the already exposed idea of the connection between closed and open shell techniques. Beside this primary reason there are others, unavoidable if general convergence, extrapolation or eigenspace manipulation devices are to be described once and for all SCF levels.
It shall be noted that each computational scheme has been tested in one way or another, and for this reason each of them can be used without further general speculation.
Without pretending to be exhaustive, some assorted examples are shown in order to give a hint on the usefulness and capability of the

procedures described herein. In this sense, molecular geometries of excited states, radicals and ions are presented, as well as some relevant properties of various open shell systems.

A final statement of completeness will be futile and, perhaps, outrageous. The authors believe in a neverending development of Science. From this scope, this work should be finally taken as an attempt to rationalize and construct the starting point of a more sophisticated and elegant SCF theory.

Chemical problems, even those attached to open shell states, are mainly related to closed shell systems. In fact, when an excitation or ionization takes place a completely new chemical structure arises. In a common closed shell species are hidden a considerable number of new structures, each one possessing its own particular chemical properties. If SCF theory has proved to be accurate enough to study closed shell problems, it hopefully can be used also to study open shell ones, particularly in cases where experimental techniques can hardly be used, and if so, only at an extraordinary cost, when compared with straightforward computations.

A theoretical structure capable of handling chemical problems must contain a set of properties which, until now, have not been reported altogether in the literature. The scattering of techniques applicable to the various parts of the SCF computational pathway is considerable, and furthermore a galaxy of open shell computational methods can be added to these.

This work should be seen as an attempt to describe a stable, general and integrated SCF theory framework, in such a manner that it can be used as well for teaching purposes, as for research and program construction.

In order to enhance this last goal a modular structure has been used throughout the description of the various SCF computational levels, although care has also been taken to establish the relationship and uniqueness of them all.

Edmonton, September, 1977.

Acknowledgements

Many people have made possible the development of a general SCF theory and the writing of this book. The authors should acknowledge the contribution of Dr. R. Caballol and Dr. R. Gallifa, the continuing work of Mr. J. A. Hernandez and Mr. F. Sanz as well as the efforts of Mr. J. Huguet, Mr. R. Carulla, Ms. D. Gibbs, Ms. N. Gilbey and Ms. J. Jorgensen. Without their help this manuscript would never have been written.
One of us (R.C.) wishes to thank Professor S. Huzinaga for the invitation during 1977, to the Department of Chemistry of the University of Alberta (Canada), making feasible the preparation of the final version of this work. The support and facilities given by the Faculty of Medicine of the Universitat Autonoma de Barcelona are also deeply acknowledged.
The authors also acknowledge the computing facilities kindly provided by Professors E. Scrocco and S. Fraga.

Acknowledgement

Contents Page

I. Historical Review

At the end of this manuscript the reader can find, as complete as possible, a bibliography on restricted Hartree-Fock and Multiconfigurational papers. Compared with the huge amount of theoretical calculations, the number of papers devoted to the theoretical and fundamental SCF framework is without doubt scarce. Table 1.I gives the number of published papers quoted in the present bibliographic survey. The total number of papers divided by the years scanned give 9 for the Open Shell and 10 for Multiconfigurational papers. Publication trends in both themes are almost the same: a painful rising through the sixties and a spectacular increase in the seventies, with particular heavy publication in the period 73-75. An also spectacular decrease can be noted in 1976, probably due to extra-scientific reasons.
The reader can extract his own conclusions and, as an extra amusement, play a forecast game on the subject.

1. The Open Shell Development

The now famous 1960 Roothaan's paper has opened the door to an interesting line of work, which is outlined in Table 2.I.
This manuscript may be put at the end of the list and his mathematical analysis can be easily related to all the quoted papers in the Table. Following Roothaan's paper, Huzinaga's 1960-61 papers began to sniff into the most difficult application part of Roothaan's formalism. From this starting point Birss and Fraga's 1963-64 work can be considered the first ambitious attempt to describe a generalized theory. The lack of the Lagrange multiplier terms in Birss and Fraga's Coupling Operator, as well as the necessary conditions of orthogonality constraints in the singlet case, has motivated without doubt Huzinaga's 1969 paper, which in many aspects contains all the requirements for further development of the theory. At the same time Goddard, Dunning and Hunt have pointed out the same facts, although in some different formalism.
The paper of Peters in 1972, retrieved the interesting advantages of a projection operator formalism in order to simplify Euler equations and precluded Hirao's development of 1974. Hirao's scheme was published practically at the same time as the present formalism, whose main papers were published in 1974 and 1975.
Hirao and Nakatsuji, however, described before the present authors, and explicitly stressed, the importance of the Lagrange multipliers hermitean conditions. In our 1974 paper we overlooked the problem,

Table 1.I

Development of the number of published papers on SCF Open Shell and Multiconfigurational theories

	Number of Papers	
Year	Open shell	Multiconfigurational
1957	1	---
1960	2	---
1961	1	---
1962	0	---
1963	2	---
1964	9	---
1965	2	1
1966	1	1
1967	3	8
1968	2	4
1969	6	6
1970	12	8
1971	10	8
1972	14	14
1973	21	21
1974	24	24
1975	32	22
1976	21	13
Total	161	130

being much more interested in the synthetic development of a general formalism, which could handle multiconfiguration open and closed shell systems. Later, reflection in the light of Hirao's paper gave as a result the 1975 work at the end of the Table 1.II. This last paper can be viewed as a final synthesis of the 1975 state of the art.

2. The Multiconfigurational Scheme

Hirao's and our group's papers, presented on Table 1.II, give the multiconfigurational structure of the SCF Coupling Operator as a product of the synthetic purposes of the authors. Once again Roothaan's 1960 work contains, in some manner, a multiconfigurational seed, although the first operative multiconfigurational computational procedures were given by Gilbert (1965) and Das and Wahl (1966).
The paper of the last authors published in 1967 is particularly interesting. In a parallel effort Hinze and Roothaan (1967) and Wessel (1967) have stated another enlightening framework; again in the same year Veillard and Clementi proposed a fruitful multiconfigurational framework, although all of them used a multiple operator scheme. It should be mentioned that the formalism developed by McWeeny in 1968 and in the same year by Levy and Berthier's paper, linked Brillouin's theorem with MC theory. In 1969 Huzinaga established an elegant formalism, contemporary to his Open Shell paper. It was not until 1973 that some significant improvement was reached: it is interesting to quote a paper of this year by Wood and Veillard, which uses the contemporary Saunders and Hillier idea of level shift technique. A very elegant and general formalism was developed also in 1973 by Hinze.

Table 2.I

Outline of the most relevant papers in the development of the Open Shell SCF Theory as formulated in the present work

C.C.J. Roothaan
Rev. Mod. Phys., 32, 179 (1960)

S. Huzinaga
Phys. Rev., 120, 866 (1969); 122, 131 (1961)

F.W. Birss, S. Fraga
J. Chem. Phys., 38, 2562 (1963); 40, 3203, 3207, 3212 (1964)

W.G. Laidlaw, F.W. Birss
Theoret. Chim. Acta, 2, 181 (1964)

S. Huzinaga
J. Chem. Phys., 51, 3971 (1969)

W.A. Goddard, T.H. Dunning, W.J. Hunt, Chem. Phys. Lett., 4, 231 (1969)

D. Peters
J. Chem. Phys., 57, 4351 (1972)

K. Hirao, H. Nakatsuji
J. Chem. Phys., 59, 1457 (1973)

R. Caballol, R. Carbo
R. Gallifa, J.M. Riera,
Int. J. Q. Chem., 8, 373 (1974)

K. Hirao
J. Chem. Phys., 60, 3215 (1974)

R. Carbo, R. Gallifa, J.M. Riera, Chem. Phys. Lett., 30, 43 (1975).

II. Electronic Energy, Fock Operators and Coupling Operators

1. Introduction

In this section the essential framework of the generalized Hartree-Fock theory will be developed. As starting point, a general energy expression and the attached operators will be discussed.
The energy expression will lead us to construct the Euler equations and the generalized Fock operators, and from here the Coupling Operator and its secular equation will be obtained.
The Coupling Operator will be divided into three well differentiated parts, two of them will be studied immediately, leaving the third for the next section.
The LCAO formulation of the previous theoretical framework will be studied next, ending the section with a derivation of simple cases, mainly monoconfigurational with open or closed shells.

2. General Energy Expression

Suppose that, following a previous computation, an electronic energy expression such as

$$E = \sum_{i,j} (\omega_{ij} h_{ij} + \sum_{k,\ell} (\alpha^{ij}_{k\ell}(ij|k\ell) - \beta^{ij}_{k\ell}(ik|j\ell)) \tag{1.II}$$

is obtained. The sums run over the intervening MO's. In any case from the SCF point of view, $\{\omega_{ij}\}$, $\{\alpha^{ij}_{k\ell}\}$ and $\{\beta^{ij}_{k\ell}\}$ are specific constants, which depend on the state wavefunction attached to the energy. They will be thereafter called <u>state parameters</u>.
The $\{h_{ij}\}$ are the core molecular integrals

$$h_{ij} = \langle i|h|j\rangle = \int \Theta^*_i(1) h(1) \Theta_j(1) dv_1,$$

calculated through the MO set $\{\Theta_i\}$ and h is the monoelectronic hamiltonian

$$h = -\frac{1}{2}\nabla^2 + \sum_A \frac{Z_A}{V_A}.$$

$\{(ij|k\ell)\}$ are, on the other hand, the electronic repulsion integrals defined as

$$(ij|k\ell) = \iint \Theta^*_i(1)\Theta^*_k(2)\; r^{-1}_{12}\; \Theta_j(1)\Theta_\ell(2)\; dv_1 dv_2;$$

it is interesting here to note that

$$(ij|k\ell) = (ji|k\ell)^* = (ij|\ell k)^* = (ji|\ell k)$$

$$= (k\ell|ij) = (k\ell|ji)^* = (\ell k|ij)^* = (\ell k|ji),$$

in this manner, the state parameters behave in a similar way, so one can write

$$\alpha_{k\ell}^{ij} = \alpha_{k\ell}^{ji^*} = \alpha_{\ell k}^{ij^*} = \alpha_{\ell k}^{ji}$$

$$= \alpha_{ij}^{k\ell} = \alpha_{ji}^{k\ell^*} = \alpha_{ij}^{\ell k^*} = \alpha_{ji}^{\ell k} \;;$$

and the same for $\beta_{k\ell}^{ij}$:

$$\beta_{k\ell}^{ij} = \beta_{kj}^{i\ell^*} = \dots ,$$

with a final relationship

$$\omega_{ij} = \omega_{ji}^* \;.$$

3. <u>General Coulomb and Exchange Operators</u>

The electron repulsion integrals can be expressed in an operator-like manner. It can be defined a set of generalized Coulomb operators $\{V_{ij}\}$, through the integrals

(2.II) $$V_{ij}(2) = \int \Theta_i^*(1)\; r_{12}^{-1}\; \Theta_j(1)\; dV_1,$$

so, the repulsion integrals can be written as

$$(ij|k\ell) = \langle k|V_{ij}|\ell\rangle = \langle i|V_{k\ell}|j\rangle.$$

Generalized Coulomb operators are hermitean:

$$\langle k|V_{ij}|\ell\rangle^* = \langle \ell|V_{ij}^+|k\rangle = \langle \ell|V_{ji}|k\rangle$$

$$= (ji|\ell k) = (ij|\ell k)^* = (ij|k\ell) = \langle k|V_{ij}|\ell\rangle$$

and can be seen as the Coulomb potential of electron (1) described by the density function $\rho(1) = \Theta_i^*(1)\ \Theta_j(1)$ acting on electron (2). In a parallel way, generalized exchange operators$\{X_{ij}\}$ can be defined in order to fulfill the integral equation

$$(3.II) \qquad X_{ij}(2)\ \Theta_k(2) = \left(\int \Theta_i^*(1)\ r_{12}^{-1}\ \Theta_k(1)\ dV_1 \right)\ \Theta_j(2),$$

that is, one can write:

$$(ij|k\ell) = \langle i|X_{j\ell}|k\rangle = \langle j|X_{ik}|\ell\rangle.$$

Then, from this it follows that there is an equivalence between Coulomb and exchange operators

$$X_{ij}(2)\ \Theta_k(2) = V_{ik}(2)\ \Theta_j(2).$$

In fact, exchange operators should be seen as a useful tool when constructing Fock operators, but it will be worthwhile to keep in mind their little, if not null, physical meaning.
Exchange operators are, as their Coulomb counterparts, also hermitean

$$\begin{aligned} \langle k|X_{ij}|\ell\rangle^* &= \langle k|V_{i\ell}|j\rangle^* \\ &= \langle j|V_{i\ell}|k\rangle \\ &= (jk|i\ell) = (i\ell|jk) \\ &= \langle \ell|X_{ij}|k\rangle. \end{aligned}$$

Using this new kind of operator, the general energy expression (1.II) can be written as

$$(4.II) \qquad E = \sum_{i,j} (\omega_{ij} h_{ij} + \sum_{k,\ell} [\alpha_{k\ell}^{ij}\ \langle i|V_{k\ell}|j\rangle - \beta_{k\ell}^{ij}\ \langle i|X_{k\ell}|j\rangle]).$$

4. Energy and Lagrangian Variation: Fock Operators and Euler Equations

The energy form, previously defined, is submitted to the current orthonormality conditions between the MO's, $\{\Theta_i\}$:

$$\langle i|j\rangle = \int \Theta_i(1)^*\Theta_j(1)\ dV_1 = \delta_{ij}.$$

Variation of the energy, in order to obtain a stationary point, should take into account the MO orthogonality. For this purpose the augmented functional

(5.II) $$L = E - \sum_{i,j} \lambda_{ij}[\langle i|j\rangle - \delta_{ij}]$$

is constructed, the set $\{\lambda_{ij}\}$ is formed by scalars and its elements are called Lagrange multipliers.
Variation of L leads to

$$\delta L = \delta E - \sum_{i,j} \lambda_{ij}[\langle \delta i|j\rangle + \langle i|\delta j\rangle],$$

which contains in turn the energy variation δE.
Variation of the energy gives

$$\delta E = \sum_{i,j} (\omega_{ij}\ \delta h_{ij} + \sum_{k,\ell} [\alpha^{ij}_{k\ell}\ \delta(ij|k\ell) - \beta^{ij}_{k\ell}\ \delta(ik|j\ell)])$$

and

*a) $$\delta h_{ij} = \langle \delta i|h|j\rangle + \langle i|h|\delta j\rangle$$

*b) $$\delta(ij|k\ell) = \delta\langle i|V_{k\ell}|j\rangle$$

$$= \langle \delta i|V_{k\ell}|j\rangle + \langle i|V_{k\ell}|\delta j\rangle + \langle i|\delta V_{k\ell}|j\rangle$$

but

$$\langle i|\delta V_{k\ell}|j\rangle = \langle \delta k|V_{ij}|\ell\rangle + \langle k|V_{ij}|\delta \ell\rangle$$

so

$$\delta(ij|k\ell) = \langle \delta i|V_{k\ell}|j\rangle + \langle i|V_{k\ell}|\delta j\rangle + \langle \delta k|V_{ij}|\ell\rangle + \langle k|V_{ij}|\delta\ell\rangle$$

*c) the same reasoning can be used for the exchange part, but using the generalized exchange operators$\{X_{ij}\}$, that is

$$\delta(ik|j\ell) = \delta\langle i|X_{k\ell}|j\rangle$$

$$= \langle \delta i|X_{k\ell}|j\rangle + \langle i|X_{k\ell}|\delta j\rangle + \langle \delta k|X_{ij}|\ell\rangle + \langle k|X_{ij}|\delta\ell\rangle.$$

In this manner δL can be divided in two well defined parts:

$$\delta L = \delta M + (\delta M)^*,$$

with

$$\delta M = \sum_{i,j} (\omega_{ij}<\delta i|h|j> + \sum_{k,\ell} [\alpha_{k\ell}^{ij}\{<\delta i|V_{k\ell}|j> + <\delta k|V_{ij}|\ell>\}$$
$$- \beta_{k\ell}^{ij}\{<\delta i|X_{k\ell}|j> + <\delta k|X_{ij}|\ell>\}])$$
$$- \sum_{i,j} \lambda_{ij}<\delta i|j>.$$

The stationary condition $\delta L = 0$ gives rise to

$$\delta M = (\delta M)^* = 0,$$

which will hold simultaneously. The first condition can be rewritten as

$$\sum_{i,j} [\omega_{ij}<\delta i|h|j> + \sum_{k,\ell} (\alpha_{k\ell}^{ij}\{<\delta i|V_{k\ell}|j> + <\delta k|V_{ij}|\ell>\}$$
$$- \beta_{k\ell}^{ij}\{<\delta i|X_{k\ell}|j> + <\delta k|X_{ij}|\ell>\})] = \sum_{i,j} \lambda_{ij}<\delta i|j>.$$

Now, using the hermitean properties of the Coulomb and exchange operators this last equation can be structured in the following form:

$$\sum_i <\delta i| \sum_j [\omega_{ij}h + 2\sum_{k,\ell} (\alpha_{k\ell}^{ij} V_{k\ell} - \beta_{k\ell}^{ij} X_{k\ell})]|j>$$
$$= \sum_i <\delta i|(\sum_j \lambda_{ij}|j>),$$

which is equivalent to the simultaneous set of equations (<u>Euler equations</u>):

(6.II) $$\sum_j [\omega_{ij}h + 2\sum_{k,\ell} (\alpha_{k\ell}^{ij} V_{k\ell} - \beta_{k\ell}^{ij} X_{k\ell})]|j> = \sum_j \lambda_{ij}|j>. \quad (\forall i)$$

At this stage, one can define the set of operators $\{F_{ij}\}$

(7.II) $$F_{ij} = \frac{1}{2}\omega_{ij}h + \sum_{k,\ell} (\alpha_{k\ell}^{ij} V_{k\ell} - \beta_{k\ell}^{ij} X_{k\ell}),$$

called <u>Fock operators</u>; in this manner the previous set of equations (6.II) can be written:

(8.II) $$\sum_j F_{ij}|j> = \sum_j \lambda_{ij}|j>; \quad \forall i,$$

where a factor 2, which should multiply the operators at the left has been considered included into the Lagrange multipliers. Beside these

equations one should take into account the conjugated ones, derived from the second variational condition $(\delta M)^* = 0$.
This will lead to:

$$(9.\text{II}) \qquad \sum_j \langle j|F_{ji} = \sum_j \lambda^*_{ji}\langle j|; \quad \forall i.$$

Using the same arguments as done previously, and taking into account the hermitean nature of the set $\{F_{ij}\}$

$$F^+_{ij} = F_{ji}; \quad \forall i,j.$$

The hermitean structure of the Fock operator matrix can be proved easily, if we take into account that

a) $\omega^*_{ij} = \omega_{ji}$

b) $(\alpha^{ij}_{k\ell})^* = \alpha^{ji}_{\ell k}$

and

c) $(\beta^{ij}_{k\ell})^* = \beta^{ji}_{\ell k}$,

as well as the Coulomb and exchange operator hermitean properties

$$F^+_{ij} = \frac{1}{2}\,\omega^*_{ij}h^+ + \sum_{k,\ell}[(\alpha^{ij}_{k\ell})^*\, V^+_{k\ell} - (\beta^{ij}_{k\ell})^*\, X^+_{k\ell}]$$

$$= \frac{1}{2}\,\omega_{ji}h + \sum_{k,\ell}(\alpha^{ji}_{\ell k}\, V_{\ell k} - \beta^{ji}_{\ell k}\, X_{\ell k}) = F_{ji}.$$

Then, the first Euler equations multiplied on the left by the bra $\langle k|$ give

$$\sum_j \langle k|F_{ij}|j\rangle = \lambda_{ik}; \quad \forall\ i,k,$$

while the conjugated equations multiplied on the right by the ket $|k\rangle$ yield

$$\sum_j \langle j|F_{ji}|k\rangle = \lambda^*_{ki}; \quad \forall\ k,i.$$

Due to the hermitean condition of Fock operators

$$\sum_j \langle k|F_{ij}|j\rangle = \sum_j (\langle k|F_{ij}|j\rangle)^* = \sum_j \langle j|F^+_{ij}|k\rangle = \sum_j \langle j|F_{ji}|k\rangle.$$

Or, one can write the following conditions on the Lagrange multipliers

$$\lambda_{ik} = \lambda^*_{ki}; \quad \forall k,i,$$

instead of the conjugated Euler equations. In other words, the matrix $\{\lambda_{ij}\}$ of Lagrange multipliers at the stationary point should be hermitean.

The Euler equations can then be written as

$$\text{(10.II)} \quad \left\{ \begin{array}{l} \sum_j F_{ij}|j\rangle = \sum_j \lambda_{ij}|j\rangle; \quad \forall i \\ \\ \text{and} \quad \lambda_{ij} = \lambda^*_{ji}; \quad \forall i,j, \end{array} \right.$$

A very general and widespread misconception on many procedures used in open shell calculations is based on ignoring the hermitean condition of Lagrange multipliers at the stationary point.

This misleading situation gives rise to many problems when $\lambda_{ij} \neq 0$. The most conspicuous one is the final SCF energy dependence on the starting vectors.

5. Coupling Operator

The term Coupling Operator will be understood here as an operator, R, constructed in such a manner that the Euler equations (10.II) can be reduced to a unique pseudosecular equation in Roothaan's sense

$$\text{(11.II)} \quad R|i\rangle = \epsilon_i|i\rangle; \quad \forall i,$$

where the term pseudosecular stands for the intrinsic dependence of R on the eigenvectors $\{|i\rangle\}$. A unique pseudosecular equation has obvious advantages over other Euler equations manipulations, which obtain as a final product a set of coupling operators each attached to one equation. The most relevant advantages of a unique operator can be listed as follows:

1) Each SCF step needs only one diagonalization
2) A unique MO set is obtained, already orthogonal if R is chosen hermitean
3) Closed shell extrapolation and other techniques developed during the past ten years can be easily used
4) Storage space is saved

A unique coupling operator R is mainly composed of three well defined parts. Two of them come from the Euler equations, and are unavoidable unless some special conditions are fulfilled; the third one can be avoided, and if not it can be composed from all the possible manipulations which leave the R occupied eigenspace invariant, so that total energy remains invariant.
Through this work the operator parts will be named as: 1) The null gradient operator, that is the only part remaining in closed shell monoconfigurational procedures. 2) Off-diagonal Lagrange multipliers hermitean conditions part, which is needed in most open shell and multiconfigurational cases, except when exceptional symmetry conditions are met. 3) Eigenvalues and eigenspace manipulation, composed by projectors and projected operators which, when used, should only leave the electronic energy expression invariant. Level shift operators and virtual eigenspace manipulation operators fall into this category. In some cases they constitute a completely arbitrary set of operators; in other situations, as in monoconfigurational excited singlet states with the same symmetry as singlet ground state, they are compulsive in order to assure proper orthogonality between the associated singlet wavefunctions, otherwise the SCF process converges to a false extremum.
The main idea which underlies the unique operator formalism is the possibility to obtain easily its spectral decomposition at self-consistency

(12.II) $$R = \sum_i \epsilon_i |i\rangle\langle i| + \sum_v \epsilon_v |v\rangle\langle v|,$$

where the first sum runs over the active MO in the energy expression, and the second over the, so called, virtual orbitals, the necessary orbitals to complete the R eigenspace.
The following discussion will develop this point for each part of the coupling operator.

6. Null Gradient Coupling Operator Part

Starting from the equation (8.II) and the value $\lambda_{ii} = \sum_j \langle i|F_{ij}|j\rangle$, the spectral decomposition

(13.II) $$R_G = \sum_i \lambda_{ii} |i\rangle\langle i|$$

gives

$$R_{G_0} = \sum_i \sum_j P_{ii} F_{ij} P_{ji}, \text{ (with } P_{ji} = |j\rangle\langle i|;\ P^+_{ji} = P_{ij})$$

which can be made hermitean through the sum

$$R_{G_1} = \frac{1}{2}(R_{G_0} + R^+_{G_0}),$$

that is

$$R_{G_1} = \frac{1}{2} \sum_i \sum_j (P_{ii} F_{ij} P_{ji} + P_{ij} F_{ji} P_{ii}).$$

this operator already serves the purpose of transforming the gradient equations into a unique pseudosecular equation, but the virtual spectrum of R_{G_1} is completely degenerate and null. In order to avoid this drawback it should be sufficient to define the projector over all virtual space $V = \{|v\rangle\}$

$$P_V = \sum_{v \in V} |v\rangle\langle v|,$$

and redefine the operator as

$$R_G = \frac{1}{2} [\sum_i \sum_j (\Pi_i F_{ij} P_{ji} + P_{ij} F_{ji} \Pi_i)$$
$$+ \sum_i \sum_j (P_{ii} F_{ij} P_V + P_V F_{ji} P_{ii})$$
$$+ P_V \{\sum_i \sum_j (F_{ij} + F_{ji})\} P_V]$$

where $\Pi_i = P_{ii} + P_V$.

Then

$$R_G|k\rangle = \frac{1}{2} \sum_j \Pi_k F_{kj}|j\rangle + \frac{1}{2} \sum_j P_{kj} F_{jk}|k\rangle$$

the first term simplifies, if one takes into account that:

$$\Pi_i|p\rangle = \delta_{ip}|p\rangle, \text{ if p occupied}$$
$$= |p\rangle, \text{ if p virtual,}$$

then:

$$R_G|k\rangle = \frac{1}{2} \lambda_{kk}|k\rangle + \frac{1}{2} \sum_j |k\rangle\langle j|F_{jk}|k\rangle$$
$$= \frac{1}{2} \lambda_{kk}|k\rangle + \frac{1}{2}(\sum_j \langle j|F_{jk}|k\rangle)|k\rangle = \lambda_{kk}|k\rangle,$$

the virtual eigenvalues of R_G are non-zero since

$$<v|R_G|v> = \frac{1}{2}\sum_i \sum_j <v|F_{ij} + F_{ji}|v>.$$

In a more general way one can define an operator

$$(14.II) \qquad R_G = \frac{1}{2}[\sum_i \alpha_i \sum_j (\Pi_i F_{ij} P_{ji} + P_{ij} F_{ji} \Pi_i) + \sum_i \sum_j (P_{ii} F_{ij} P_V + P_V F_{ji} P_{ii}) + P_V \sum_i \{\sum_j (F_{ij} + F_{ji})\} P_V]$$

where $\{\alpha_i\}$ is a set of arbitrary real parameters, then

$$R_G|k> = \epsilon_k|k>; \quad \forall k,$$

with

$$\epsilon_k = \alpha_k \lambda_{kk}; \quad \forall k.$$

7. Lagrange Multipliers Hermitean Condition Coupling Operator Part

It has been shown that the Lagrange multipliers can be written as

$$\lambda_{ik} = \sum_j <k|F_{ij}|j>; \forall i,k,$$

then, also

$$\lambda_{ki} = \sum_j <i|F_{kj}|j>; \quad \forall i,k,$$

taking the conjugate of this last equality

$$\lambda^*_{ki} = \sum_j <j|F^+_{kj}|i> = \sum_j <j|F_{jk}|i>,$$

and the hermitean conditions of Lagrange multipliers it can be expressed as

$$\lambda_{ik} - \lambda^*_{ki} = \sum_j (<k|F_{ij}|j> - <j|F_{jk}|i>) = 0; \quad \forall i,k.$$

An operator can be constructed now within the whole MO set, by means of

$$(15.II) \qquad R_{H_O} = \sum_p \sum_q \mu_{pq}|p><q|,$$

the matrix representation of R_{H_O} in the basis of the MO set is

$$<i|R_{H_O}|k> = \mu_{ik},$$

then, one can choose

$$\mu_{ik} = \rho_{ik}(\lambda_{ik} - \lambda^*_{ki}),$$

in principle, $\{\rho_{ik}\}$ being arbitrary scalars

$$R_{H_O} = \sum_p \sum_q \rho_{pq} \sum_j (<p|F_{qj}|j> - <j|F_{jp}|q>)|p><q|$$

or

$$R_{H_O} = \sum_p \sum_q \rho_{pq} \sum_j (P_{pp} F_{qj} P_{jq} - P_{pj} F_{jp} P_{qq}).$$

For reasons of coherence, R_{H_O} can be made hermitean, using

$$R^+_{H_O} = \sum_p \sum_q \rho^*_{pq} \sum_j (P_{qj} F_{jq} P_{pp} - P_{qq} F_{pj} P_{jp}).$$

Changing the role of p and q

$$R^+_{H_O} = \sum_p \sum_q \rho^*_{qp} \sum_j (P_{pj} F_{jp} P_{qq} - P_{pp} F_{qj} P_{jq}),$$

which can be written as

$$R^+_{H_O} = - \sum_p \sum_q \rho^*_{qp} \sum_j (P_{pp} F_{qj} P_{jq} - P_{pj} F_{jp} P_{qq}),$$

and the hermitean operator

$$R_H = R_{H_O} + R^+_{H_O}$$

constructed as

(16.II) $$R_H = \sum_p \sum_q \sigma_{pq} \sum_j (P_{pp} F_{qj} P_{jq} - P_{pj} F_{jp} P_{qq})$$

where

$$\sigma_{pq} = \rho_{pq} - \rho^*_{qp}$$

are the elements of a skew hermitean matrix

$$\sigma^*_{qp} = \rho^*_{qp} - \rho_{pq} = - \sigma_{pq}.$$

The matrix elements of R_H with respect to the molecular basis are a synthesis of the Lagrange multipliers hermitean conditions:

$$\text{(17.II)} \quad \langle i|R_H|k\rangle = \sum_p \sum_q \sum_j \sigma_{pq}\{\delta_{ip}\ \delta_{kq}\}[\langle p|F_{qj}|j\rangle - \langle j|F_{jp}|q\rangle]$$

$$= \sigma_{ik} \sum_j [\langle i|F_{kj}|j\rangle - \langle j|F_{ji}|k\rangle]$$

$$= \sigma_{ik}(\lambda_{ki} - \lambda^*_{ik})$$

provided that

$$\langle i|R_H|k\rangle = 0; \quad \forall i,k.$$

and $\qquad \sigma_{ik} \neq 0.$

8. LCAO Form of Coupling Operator

In this section the LCAO structure of the Fock operators, Coupling operator and the attached pseudosecular equations will be given. As usual we start from a set of AO's $\{\chi_\mu\}$ and construct the MO, used throughout the previous discussion, as linear combinations of the atomic basis set

$$|i\rangle = \sum_\mu C_{i\mu}\ \chi_\mu,$$

$\{C_{i\mu}\}$ being the coordinates of MO $|i\rangle$ with respect to the AO basis set. The set of these coordinates will be expressed as the set of column vectors $\{C_i\}$.

The matrix representation of the mentioned operators and equations follows:

a) Fock Operators:

Taking equation (7.II) as the form of the set $\{F_{ij}\}$ of Fock operators, then:

$$F_{ij,\mu\nu} = \frac{1}{2}\ \omega_{ij} h_{\mu\nu} + \sum_{k,\ell} [\alpha^{ij}_{k\ell} \sum_{\lambda,\sigma} C_{k\lambda} C^*_{\ell\sigma}(\mu\nu|\lambda\sigma) - \beta^{ij}_{k\ell} \sum_{\lambda,\sigma} C_{k\lambda} C^*_{\ell\sigma}(\mu\lambda|\nu\sigma)],$$

and rearranging terms

$$F_{ij,\mu\nu} = \frac{1}{2}\ \omega_{ij} h_{\mu\nu} + \sum_{\lambda,\sigma} [A^{ij}_{\lambda\sigma}(\mu\nu|\lambda\sigma) - B^{ij}_{\lambda\sigma}(\mu\lambda|\nu\sigma)],$$

where:

$$h_{\mu\nu} = \int \chi_\mu^*(1)\ h\ \chi_\nu(1)\ dv_1$$

$$(\mu\nu|\lambda\sigma) = \int \chi_\mu^*(1)\ \chi_\lambda^*(2)\ r_{12}^{-1}\ \chi_\nu(1)\ \chi_\sigma(2)\ dv_1 dv_2,$$

and

$$A_{\lambda\sigma}^{ij} = \sum_{k,\ell} \alpha_{k\ell}^{ij}\ C_{k\ell}\ C_{\ell\sigma}^*$$

$$B_{\lambda\sigma}^{ij} = \sum_{k,\ell} \beta_{k\ell}^{ij}\ C_{k\lambda}\ C_{\ell\sigma}^*.$$

b) Coupling Operator

Let

$$P_{ij} = C_i\ C_j^+,$$

then, the set of projectors $\{P_{ij}\}$ have the matrix representation:

$$|i\rangle\langle j| = C_i\ C_j^+\ S = P_{ij}S$$

if used as right projectors, with $S_{\mu\nu} = \int \chi_\mu^*(1)\ \chi_\nu(1)\ dv_1$, as the metric elements.

With this definition, the operator R_G in (14.II) can be written in a first step with the aid of

$$R_0 = \frac{1}{2}\ [\sum_i \sum_j (\Pi_i\ F_{ij}\ P_{ji} + P_{ij}\ F_{ji}\ \Pi_i) + \sum_i \sum_j (P_{ii}\ F_{ij}\ P_V + P_V\ F_{ji}\ P_{ii}) + P_V (\sum_i \sum_j (F_{ij} + F_{ji}))\ P_V] \quad (18.II)$$

and the true operator will be finally:

$$R = S\ R_0 S\ . \quad (19.II)$$

c) Pseudosecular Equation

The matrix form of the coupling operator pseudosecular equation can be written as:

$$R\ C_i = \varepsilon_i S\ C_i,$$

or

$$R_0 S\ C_i = \varepsilon_i\ C_i.$$

Two procedures can be used to solve this generalized secular equation:

Cholesky's method

$$S = T^{+}T \text{ with } T \text{ being triangular superior matrix,}$$

then $R_O T^{+} T\, C_i = \varepsilon_i C_i$,

and also

$$TR_O T^{+} TC_i = \varepsilon_i C_i .$$

Calling $d_i = TC_i$

and $R_T = TR_O T^{+} \rightarrow R_T^{+} = R_T$,

so

$$R_T d_i = \varepsilon_i d_i \rightarrow d_i^{+} d_j = \delta_{ij}$$

and $C_i = T^{-1} d_i$.

Löwdin's method

If, using the positive definite metric matrix S, a square root decomposition is envisaged

$$S = S^{+1/2}\, S^{+1/2},$$

through the secular equation of S

$$SD = D\Sigma,$$

and

$$S^{\pm 1/2} = D\, \Sigma^{\pm 1/2}\, D^{+},$$

then, the transformation of R_O

$$R_L = S^{+1/2}\, R_O\, S^{+1/2}$$

gives

$$R_L d_i = \varepsilon_i d_i$$

with

$$c_i = S^{-1/2}\, d_i .$$

From the computational point of view Cholesky's method is more accurate and its speed far greater than Löwdin's method.

9. Simplified Energy Forms. Monoconfigurational Open Shell

In the most current monoconfigurational cases the electronic energy and hence the attached Fock operators, Euler equations and final Coupling operator forms become simpler than the SCF structure already studied. If in the energy expression the state parameters have the following properties

1) $$\alpha^{ij}_{k\ell} = \alpha_{ij}\, \delta_{k\ell}$$

$$\beta^{ij}_{k\ell} = \beta_{ij}\, \delta_{k\ell}$$

$$\omega_{ij} = \delta_{ij}\omega_i$$

the final energy form will be

(20.II) $$E = \sum_i (\omega_i h_{ii} + \sum_j (\alpha_{ij} J_{ij} - \beta_{ij} K_{ij})),$$

$\{J_{ij}\}$ being the Coulomb integrals $(ii|jj)$ and $\{K_{ij}\}$ the exchange integrals $(ij|ij)$.

Using the operators $J_j = V_{jj}$ and $K_j = X_{jj}$,

(21.II) $$E = \sum_i (\omega_i h_{ii} + \sum_j (\alpha_{ij} (i|J_j|j) - \beta_{ij} (i|K_j|i))),$$

defining the Fock operators

(22.II) $$F_i = \frac{1}{2}\, \omega_i h + \sum_j (\alpha_{ij} J_j - \beta_{ij} K_j),$$

then, the Euler equations become

(23.II) $$\begin{cases} \text{a)} & F_i|i\rangle = \sum_j \lambda_{ij}|j\rangle, \ \forall\, i \\ \text{b)} & \lambda_{ij} = \lambda^*_{ji}, \ \forall\, i,j. \end{cases}$$

The final form of the coupling operator in this case can be easily obtained, taking into account that for simplification purposes, one can consider:

$$F_{ij} = \delta_{ij} F_i,$$

then

a) The <u>Gradient part of the operator</u> can be written:

$$\begin{aligned} R_G &= \frac{1}{2} \sum_i \sum_j \delta_{ij} (\Pi_i F_i P_{ji} + P_{ij} F_i \Pi_i) \\ &+ \frac{1}{2} \sum_i \sum_j \delta_{ij} (P_{ii} F_i P_V + P_V F_i P_{ii}) \\ &+ \frac{1}{2} P_V (\sum_i \sum_j \delta_{ij} (2F_i)) P_V \\ &= \frac{1}{2} \sum_i (\Pi_i F_i P_{ii} + P_{ii} F_i \Pi_i) \\ &+ \frac{1}{2} \sum_i (P_{ii} F_i P_V + P_V F_i P_{ii}) \\ &+ P_V (\sum_i F_i) P_V \end{aligned}$$

or

$$\begin{aligned} R_G &= \frac{1}{2} \sum_i (P_{ii} F_i P_{ii} + P_{ii} F_i P_{ii}) \\ &+ \frac{1}{2} \sum_i (P_V F_i P_{ii} + P_{ii} F_i P_V) \\ &+ \frac{1}{2} \sum_i (P_{ii} F_i P_V + P_V F_i P_{ii}) \\ &+ P_V (\sum_i F_i) P_V \end{aligned}$$

and finally

$$R_G = \sum_i (P_{ii} F_i P_{ii} + P_V F_i P_{ii} + P_{ii} F_i P_V + P_V F_i P_V),$$

in this manner it can be written

(24.II) $$R_G = \sum_i \Pi_i \, F_i \, \Pi_i,$$

which is the final gradient form of the operator.
Also, using an arbitrary set of real numbers $\{\alpha_i\}$,

(25.II) $$R_G = \sum_i \alpha_i \, \Pi_i \, F_i \, \Pi_i$$

b) The Lagrange multipliers hermitean conditions part of the coupling operator can be written as:

$$R_H = \sum_p \sum_q \sigma_{pq} \sum_j (\delta_{qj} P_{pp} F_j P_{jq} - \delta_{jp} P_{pj} F_j P_{qq})$$

$$= \sum_p \sum_q \sigma_{pq} P_{pp} F_q P_{qq} - \sum_p \sum_q \sigma_{pq} P_{pp} F_p P_{qq},$$

that is

(26.II) $$R_H = \sum_p \sum_q \sigma_{pq} P_{pp} (F_q - F_p)\, P_{qq}$$

the set of parameters $\{\sigma_{pq}\}$ conserving the same meaning as in the previous definition.

10. Closed Shell

From the simplified energy form, it is easy to obtain the closed shell SCF computational structure. In this situation:

$$\alpha_{ij} = 2 \quad \forall\, i,j,$$

$$\beta_{ij} = 1 \quad \forall\, i,j,$$

and $$\omega_i = 2 \quad \forall\, i,$$

so the energy is

(27.II) $$E = \sum_i (2\, h_{ii} + \sum_j (2\, J_{ij} - K_{ij})),$$

consequently:

(28.II) $$F_i = F = h + \sum_j (2\, J_j - K_j); \quad \forall\ i,$$

then, the gradient part of the coupling operator is

(29.II) $$R_G = \sum_i \Pi_i\, F\, \Pi_i$$

$$= \sum_i (P_i + P_V)\, F (P_i + P_V)\,.$$

This expression, from the occupied MO point of view, is equivalent to

$$R_G = F;$$

use of

$$R_G = \sum_i \alpha_i (\Pi_i \; F \; \Pi_i)$$

will only multiply the F eigenvalues by the $\{\alpha_i\}$ factors.
Since the differences between Fock operators of each shell vanish, in this case:

$$R_H = 0.$$

11. Corollary

From a general energy expression, a unique pseudosecular equation attached to a unique coupling operator can be obtained, upon variation. This last form can be applied without further manipulations to any simpler case.
The Coupling Operator is constructed in such a manner that it can be used to obtain the initial energy expression. This is easily verified by taking into account that at self-consistent stage

$$\begin{aligned}
<p|R|q> &= <p|R_G|p>\delta_{pq} \\
&= \frac{1}{2} \sum_i <p|\sum_j \Pi_i \; F_{ij} \; P_{ji} + P_{ij} \; F_{ji} \; \Pi_i|p> \\
&= \frac{1}{2} \sum_j \{<p|F_{pj}|j> + <j|F_{jp}|p>\} \\
&= \sum_j \{\frac{1}{2} \omega_{pj} \; h_{pj} + \sum_{k,\ell} [\alpha_{k\ell}^{pj} \; (pj|k\ell) - \beta_k^{pj} \; (p\ell|jk)]\}
\end{aligned}$$

then summing over p and q and rearranging the indexes

$$\sum_p \sum_q <p|R|q> = \sum_p \sum_q \{\frac{1}{2} \omega_{pq} \; h_{pq} + \sum_{k,\ell} [\alpha_{k\ell}^{pq} (pq|k\ell) - \beta_{k\ell}^{pq} (p\ell|qk)]\}$$

so if one adds to the R operator the terms $\frac{1}{2} \omega_{pq}$ h:

$$E = \sum_p \sum_q <p|R + \frac{1}{2} \omega_{pq} \; h|q> , \qquad (30.II)$$

which is a very well known expression in the monoconfigurational closed shell cases, where R = F.

III. Eigenspace Manipulations

1. Introduction

Although the formal results of this section can have multiple applications, they were originally studied for application to SCF theory, and even in their original form they can be used in the framework of a unique coupling operator formalism.
In order to develop a useful theoretical background with immediate application to a computational algorithm, the most convenient procedure is to start with a generalized secular equation in matrix form

$$(1.\text{III})\qquad R_0|0;i\rangle = \lambda_{0;i}\, S|0;i\rangle,\; i \in N$$

with $R_0^+ = R_0, S^+ = S$ the positive definite metric matrix, and the eigenvectors are orthonormalized in the sense

$$\langle i;0|S|0;j\rangle = \delta_{ij}.$$

The hermitean matrix R_0 can be modified by a new hermitean matrix V, to form

$$(2.\text{III})\qquad R = R_0 + V\,.$$

In order to observe the action of V in the eigenspace of R_0, it is necessary to consider V as the projection of an hermitean arbitrary matrix, say M, over some subspace of the eigenspace of R_0, that is

$$(3.\text{III})\qquad V = Q^+ M\, Q,$$

so the matrix Q should be regarded as a right eigenprojector over such subspace, whose eigenvectors can be attached to some subindex set $P \subset N$, then

$$(4.\text{III})\qquad Q = \sum_{p\in P} |0;p\rangle\langle p;0|S$$

and, at the same time

$$Q|0;i\rangle = \sum_{p\in P} |0;p\rangle\langle p;0|S|0;i\rangle = \sum_{p\in P} \delta_{ip}|0;p\rangle.$$

So, if $i\in P$ then

$$(5.\text{III})\qquad Q|0;i\rangle = |0;i\rangle$$

and if $i \notin P$ then

(5'.III) $Q|0;i\rangle = 0$.

No restriction is necessary on the nature of the subset of vectors with indices P. For example, they can be the virtual orbitals or any other set.
When the images of $\{|0;i\rangle\}$ under the action of R are searched for, the following is obtained

(6.III) $R|0;i\rangle = (R_0 + V)|0;i\rangle = \lambda_{0;i} S|0;i\rangle = \sum_{p\varepsilon P} \delta_{ip} Q^+ M|0;p\rangle$,

or using the definition of Q

$$R|0;i\rangle = \lambda_{0;i} S|0;i\rangle + \sum_{p\varepsilon P} \sum_{q\varepsilon P} \delta_{ip} \langle q;0|M|0;p\rangle \; S|0;q\rangle .$$

The straightforward result is that if $i \notin P$, the generalized secular equations for R and R_0 are the same, but when $i \varepsilon P$

(7.III) $R|0;i\rangle = \lambda_{0;i} \; S|0;i\rangle + \sum_{q\varepsilon P} \langle q;0|M|0;i\rangle \; S|0;q\rangle$.

In the simplest case when P has only one element, then only one modified equation should be associated to this element, say the ith equation, so one can write:

(8.III) $R|0;i\rangle = (\lambda_{0;i} + \langle i;0|M|0;i\rangle) \; S|0;i\rangle$,

and the eigenvectors of R remain invariant, but the ith eigenvalue is modified by the value $\langle i;0|M|0;i\rangle$.
This property can be related to deflation techniques in the diagonalization problem of generalized secular equations. Also the level shift technique, which is the main final objective of this discussion, is based primarily on this result.
Another useful example is

$$M = - \lambda_{0;i} \; S,$$

then

$$M|0;i\rangle = 0 \; S|0;i\rangle .$$

There are, however, other possibilities: a degeneracy pattern can be

removed or created in the spectra. For example, a particular eigenvalue, say $\lambda_{o;i}$ can be shifted in such a way that it now coincides with one of the other eigenvalues, say $\lambda_{o;k}$.
This can be realized if

$$\langle i;O|M|O;i\rangle = \lambda_{o;k} - \lambda_{o;i}.$$

In this manner the state $|O;k\rangle$ has an "accidental" degeneracy and an arbitrariness is introduced in the eigenvector space, since any linear combination of $|O;i\rangle$ and $|O;k\rangle$ can be used as an eigenvector of R.
Furthermore, a similar but more general situation can happen even when P consists of two or more indices, $\{i\}\varepsilon$ P. In this case

$$\text{(9.III)} \quad R|O;i\rangle = (\lambda_{o;i} + \langle i;O|M|O;i\rangle)\, S|O;i\rangle + \sum_{\substack{q\varepsilon P \\ (q\neq i)}} \langle q;O|M|O;i\rangle\, S|O;q\rangle; \quad i\varepsilon P.$$

As a special situation of this result let us suppose that the matrix M is diagonal with respect to the vectors $\{|O;p\rangle\}$; $p\varepsilon P$, or

$$\langle p;O|M|O;q\rangle = \mu_p\, \delta_{pq},$$

where $\{\mu_p\}$ are any numbers. Then, one has

$$V = Q^{+}MQ = \sum_{p\varepsilon P} \mu_p\, S|O;p\rangle\langle p;O|S,$$

so

$$R|O;i\rangle = (\lambda_{o;i} + \mu_i)\, S|O;i\rangle; \quad i\varepsilon P.$$

The implications become more apparent if we consider the case where all eigenvalues have a common shifting parameter, say, $\lambda_{o;v}$

$$\mu_i = \lambda_{o;v} - \lambda_{o;i}$$

Then, the eigenvalue $\lambda_{o,v}$ has now a manyfold degeneracy and the eigenvectors have an accompanying arbitrariness.
Now, going back to the general case, where the matrix M is not necessarily diagonal with respect to the vectors $\{|O;p\rangle\}$, $p\varepsilon P$, and considering a new matrix T, with elements defined according to the expression

(10.III) $t_{ij} = \delta_{ij}\lambda_{o;i} + \langle i;o|M|o;j\rangle.$

The matrix T will be hermitean, and its dimensions will depend on the number of P elements. In this manner, equation (9.III) can be written as

(11.III) $R|o;i\rangle = \sum_{q\varepsilon P} t_{qi}\, S|o;q\rangle.$

It is easy to see that R will be an endomorphism in the vector subspace of basis $\{|o;i\rangle\}$, $i\varepsilon P$, because it transforms the basis set into a set of vectors of this space.
Thus, the generalized eigenvalue equations of R will be

(12.III) $R|i\rangle = \lambda_i\, S|i\rangle; \quad i\varepsilon N,$

noting that

(13.III) if $i\varepsilon P$: $\lambda_i = \lambda_{o;i}$ and $|i\rangle = |o;i\rangle$

(13'.III) or $i\not\varepsilon P$: $\lambda_i \neq \lambda_{o;i}$ and $|i\rangle = \sum_{p\varepsilon P} a_{pi}|o;p\rangle.$

The last equations are obtained by utilizing the linear independence of the basis set $\{|o;i\rangle\}$, $i\varepsilon P$, and the endomorphic nature of R. Substituting equation (13.III) into equation (12.III), for all indices $i\varepsilon P$, it is found

(14.III) $R|i\rangle = \sum_{p\varepsilon P} a_{pi}\, R|o;p\rangle = \lambda_i \sum_{q\varepsilon P} a_{qi}\, S|o;q\rangle.$

Use of equation (11.III) and multiplication from the left by S^{-1} brings equation (14.III) to

(15.III) $\sum_{p\varepsilon P} a_{pi} \sum_{q\varepsilon P} t_{qp}|o;q\rangle = \lambda_i \sum_{q\varepsilon P} a_{qi}|o;q\rangle.$

By arrangement of equation (15.III)

(16.III) $\sum_{q\varepsilon P} \left(\sum_{p\varepsilon P} a_{pi}t_{qp}\right)|o;q\rangle = \lambda_i \sum_{q\varepsilon P} a_{qi}|o;q\rangle$

and then equating the coefficients in equation (16.III), the following is obtained

(17.III) $\sum_{p\varepsilon P} t_{qp}\, a_{pi} = \lambda_i\, a_{qi}; \quad q,i\varepsilon P.$

In a compact matrix form equation (17.III) may be written as

(18.III) $T\ A = A\ \Lambda,$

with obvious definitions for matrices T, A and Λ. Equation (18.III) is a new secular equation with a spectrum equal to the modified eigenvalues of R, and A is a unitary matrix whose columns, being the T eigenvectors, are also the coefficients of the linear combinations (13.III); from them can be obtained each modified eigenvector of R. Neither normalization nor orthogonalization needs to be carried out, that is

$$\begin{aligned} \langle i|S|j\rangle &= \sum_{p\varepsilon P} \sum_{q\varepsilon P} a^*_{pi}\ a_{qj}\ \langle p;0|S|0;q\rangle \\ &= \sum_{p\varepsilon P} \sum_{q\varepsilon P} a^*_{pi}\ a_{qj}\ \delta_{pq} \\ &= \sum_{p\varepsilon P} a^*_{pi}\ a_{pj} \\ &= \delta_{ij}\ . \end{aligned}$$

The last equality results from the unitary structure of $A(A^+A = I)$. The practical conclusion is that instead of solving the secular system (12.III) to obtain the modified vectors, the problem can be reduced to the construction of a hermitean matrix T and a further diagonalization of lower order than the original one. The reason for this general structure is that if the metric in equation (1.III) is unit (S = I), the foregoing treatment is still valid, so it will not be necessary to separate a full overlap from a ZDO formalism.

2. General Formalism

Let's come back again to equation (1.III). From the R_0 eigenspace $\mathscr{V}$, it is always possible to obtain some separate mutually exclusive linear subspaces $\{\mathscr{V}_k\}$, $k \varepsilon M$, which fulfill

$$\mathscr{U} = \bigoplus_k \mathscr{V}_k \subseteq \mathscr{V}.$$

Also, in each subspace $\mathscr{V}_k$ one can choose a basis set, formed by some subset of R_0 eigenvectors $\{|0;p\rangle\}$ $(p\varepsilon N_k)$, with the index sets N_k, being defined through the property:

$$\bigcup_k N_k \subseteq N.$$

In this manner, for each $\mathcal{V}_k$ can be defined right and left projectors P_k and P_k^+, respectively, as

$$P_k = \sum_{p \varepsilon N_k} |0;p\rangle\langle p;0|S \quad ; \; k \; \varepsilon \; M.$$

The initial matrix R_0 can be modified with a set of arbitrary hermitean matrices $\{A_k\}$, projected into each subspace $\mathcal{V}_k$; that is, we can define a new hermitean matrix, which is a generalized version of the matrix defined in equation (2.III):

$$\text{(19.III)} \quad R = R_0 + \sum_{k \varepsilon M} P_k^+ A_k P_k .$$

The behaviour of the new matrix, with respect to the characteristic basis of the initial one, provides the way to generalize the previous results.
If the secular equation of R is written in the compact form

$$\text{(20.III)} \quad R\,C = S\,C\,\Lambda$$

where the columns of C are the new eigenvectors $\{|i\rangle\}$ ($i \varepsilon N$), and $\Lambda = \mathrm{Diag}(\lambda_i,\; i \varepsilon N)$ contains the new eigenvalues, then it is possible to write

$$\text{(21.III)} \quad C = C_0 D,$$

where C_0 has as columns the R_0 eigenvectors and D is a nonsingular matrix to be determined. Substituting (21.III) in (20.III), and multiplying on the left by C_0^+, it is found

$$\text{(22.III)} \quad B\,D = D\,\Lambda,$$

where

$$B = C_0^+ \, R \, C_0$$

and use is made of the generalized orthonormality conditions

$$I = C_0^+ \, S \, C_0 \, .$$

Matrix B has a block diagonal structure: all the columns of C_0, which do not enter in the $\{\mathcal{V}_k\}$ definition, contribute to the formation of a diagonal matrix:

$$B_O = \mathrm{Diag}(\lambda_{o;i};\ i \in \bigcup_k N_k),$$

the remaining columns form a set of matrices $\{B_k\}$, whose elements are defined by

$$(23.\mathrm{III}) \quad b_{k,pq} = \lambda_{o;p}\ \delta_{pq} + \langle p;O|A_k|O;q\rangle .$$

It is easy to see then that D has also a block diagonal structure, whose elements are the unit matrix, and a set $\{D_k\}$ which can be calculated with the aid of the set of secular equations

$$B_k D_k = D_k \Lambda_k;\ k \in M$$

with

$$\Lambda_k = \mathrm{Diag}(\lambda_i;\ i \in N_k).$$

As a consequence, part of the spectrum of R_O and R is common. The modified spectral part is commanded by the nature of each $\mathscr{V}_k$ space and each A_k matrix.
Moreover, $\{A_k\}$ being hermitean, definition (21.III) gives as a result $\{B_k\}$ hermitean, hence $\{D_k\}$ is a set of unitary matrices, and D is unitary, this is consistent with the orthonormality conditions of the generalized eigenvectors of R,

$$C^+ S C = D^+ C_O^+ S C_O D = D^+ D = I.$$

In order that the treatment be more general, the space $\mathscr{U}$ can be defined, not as a direct sum as at the beginning of this discussion, but by means of a union of the space set $\{\mathscr{V}_k\}$ or

$$\mathscr{U} = \bigcup_k \mathscr{V}_k \subseteq \mathscr{V}$$

so, equation (22.III) is not fully separable, but splits into the diagonal matrix B_O and the matrix B_1, whose components are defined, in general, as

$$(24.\mathrm{III}) \quad b_{1,pq} = \lambda_{o;p}\ \delta_{pq} + \langle p;o|\sum_k A_k|o;q\rangle .$$

he sum in (24.III) extends over all the indices, whose associated spaces $\mathscr{V}_k$, fulfill

$$\{|0;p\rangle;\ |0;q\rangle\} \subset \bigcup_k \mathscr{V}_k .$$

The structure of the problem takes, in this case, a computational form similar to the previous initial treatment; they differ in that the possibility is open to use various matrices, which partition their influence among the various sets of eigenvectors.

3. Unconditional Convergence in SCF Procedures: Level Shift Techniques

It is well known that Hartree-Fock procedures do not converge in some cases. Doubtless, a good deal of computation time has been lost due to this remarkable question, but few solutions have been published. The work of Saunders and Hillier seems to be, to date, the only way to cope with the problem. Hillier and Saunders procedure has been proved convenient in many calculations and it can be used in closed or open shell computational algorithms.

3.1. The Ordering Principle

Although there can be other causes, nonconvergence in SCF calculations is mainly due to the mixing among MO's belonging to the various subsets of the molecular basis, as long as the iterative SCF path is followed. A simple traceback analysis of any reluctant case can make this clear. In the original level shift technique, the main purpose is to separate the eigenvalues of each MO within each subset (for example: the occupied MO and the virtual orbitals), preserving the initial ordering of the orbital eigenvalues so the SCF iterations, through an invariant Aufbau principle, converge to an "a priori" chosen wavefunction.
Then, a SCF desideratum can be transformed into a principle to be followed by any potential user of the Level Shift Technique: "Given an initial ordered set of MO's, the ordering and symmetry of each MO subset shall be maintained until self-consistency is achieved".
The ordering principle is a sine-qua-non condition which, if not fulfilled, brings about the result that the Aufbau principle cannot be invariant throughout SCF iterations. As a consequence, energy fluctuates between two different values and in some cases MO symmetry is lost.

3.2. Level Shift Technique in a General SCF Procedure

The principal drawback of many SCF formalisms consists of the definition of a pseudosecular equation for each shell. The Level Shift Technique has been described up to now for these kinds of formalisms.
It has been proved previously that a general procedure can be built-up, within a unique formalism, for closed, open and MC problems, which brings any SCF calculation to an unique pseudosecular equation. But

from this equation the ordering principle does not necessarily follow. It is interesting to note here that for cases in which, for some formalisms, the SCF process diverges (for instance, the OH radical in Roothaan's formalism as quoted by Sleeman), a smooth convergence is found using a unique equation.

Other cases diverge in any formalism and the principal examples of this can be found in molecular structures very far from equilibrium, where the tendency of orbital reordering is very strong.

In every case the way to follow the ordering principle is through previous experience in eigenspace manipulation of SCF solutions. The shift operator can be defined as

$$(25.\text{III}) \quad V = \sum_i \beta_i |i\rangle\langle i|$$

by means of the coupling operator R_0 eigenspace projectors $\{|i\rangle\langle i|\}$ and an arbitrary set of real numbers $\{\beta_i\}$. A new coupling operator, written as

$$(26.\text{III}) \quad R = R_0 + V$$

has the same set of eigenvectors as R_0, with a secular equation

$$(27.\text{III}) \quad R|i\rangle = (\lambda_i + \beta_i)|i\rangle,$$

that is with eigenvalues $\{\lambda_i + \beta_i\}$. As a consequence, the electronic energy, associated to the MO set $\{|i\rangle\}$ remains invariant. A complete separation of each MO, or entire subsets of the MO manifold, can be easily obtained choosing the appropriate shifts.

In this way the ordering principle can be fulfilled, and one can obtain any desired wavefunction. From this basic property of the coupling operator R, there can be constructed a flexible algorithm, which can be adapted to each case, even at each iteration, without extra effort. In this context, Saunders and Hillier Level Shift Technique can be made equivalent to a level shift operator written as

$$(28.\text{III}) \quad V = \sum_L \beta_L \Big(\sum_{i \varepsilon L} |i_L\rangle\langle i| \Big)$$

the first sum running over some MO subsets previously defined within each case.

Finally, it should be said that the ordering principle is essential not only for the general SCF computational behaviour, but also for the application of acceleration or extrapolation procedures.

The LCAO form of an operator as used in equation (25.III) will be simply a matrix defined by

$$(29.\mathrm{III}) \quad V = S \left[\sum_i \beta_i C_i C_i^+\right] S,$$

where $\{C_i\}$ are the MO expansion coefficients and S the metric matrix.

3.3 Koopman's Theorem and Level Shifted Closed Shell Fock Operators

Koopman's Theorem has been a celebrated and widely used argument in order to have an estimate of ionization potentials for closed shell molecules.

In this case, use of level shifted closed shell Fock operators written as

$$F = F_0 + \sum_i \beta_i |i\rangle\langle i|$$

permits one to obtain, through the usual SCF procedure, the proper canonical orbitals. It is evident that all the mathematical facts which apply to F_0 can be used on F. Then, using Koopman's result

$$I_i = - \langle i|F|i\rangle = - (\lambda_i + \beta_i)$$

or in this context using the arbitrariness of β_i we can have an ionization potential tailor-made. We feel that this can add more information on recently quoted "breakdowns" and inconsistencies of the theorem. We have shown that an arbitrary reordering of MO's can be obtained with the level shift technique. In this manner one can equate the MO energies to experimental spectra, and as a consequence one can safely conclude that, even in a closed shell level, MO eigenvalues have little physical meaning.

A reflection on this result should be made. Since MO energies are not observables, then any discussion involving them is, in general, meaningless. Ionization potentials should be associated with energy differences between ionized and neutral states.

3.4 Final Remarks

The values of level shifts $\{\beta_i\}$ can have a wide range of variation. In some cases relatively high values are needed in order to achieve convergence. But the higher the $\{\beta_i\}$ values, the slower convergence will be.

In any case it is worthwhile to decrease shift values at each iteration, although any function can be used.

$$\beta_i' = \beta_i(\text{iteration number})^{-x}$$

has proven to be a useful choice. The exponent x gives the rate of decrement of the shifts. Typical values ranging from x = 1.5 to x = 2 have been convenient for most cases.

IV. Multiconfigurational Structure of Monoconfigurational SCF Procedures

In light of recent discussions on the solution of the Hartree-Fock equations, the problem can be considered clarified from a formal and practical point of view in the monoconfigurational case, but in a multiconfigurational framework some of the proposed coupling operators have not yet been thoroughly discussed.
The purpose of the following sections is twofold. First, we intend to analyze the monoconfigurational problem from a new point of view. Second , using the results of this previous step, we will study the behaviour and adequacy of the coupling operator formalism, which is not yet proved to be suitable for computational purposes.

1. Monoconfigurational Energy and Euler Equations

The energy of a polyelectronic system, described by a restricted monoconfigurational wavefunction, may be written as

$$(1.IV)\quad E = \sum_i \omega_i h_{ii} + \sum_i \sum_j (\alpha_{ij} J_{ij} - \beta_{ij} K_{ij}),$$

where i, j are indexes belonging to the set $\mathcal{M}_o$, which identifies the MO's; $\{\omega_i\}$, $\{\alpha_{ij}\}$, $\{\beta_{ij}\}$ are the state parameters of the system, and $\{h_{ij}\}$, $\{J_{ij}\}$, $\{K_{ij}\}$ are the well-known core, Coulomb and exchange integrals respectively. A variation of equation (1.IV) gives the Euler equations

$$(2a.IV)\quad F_i|i\rangle = \sum_j \lambda_{ji}|j\rangle; \qquad i \in \mathcal{M}_o$$

$$(2b.IV)\quad \lambda_{ji} = \lambda^*_{ij}; \qquad i, j \in \mathcal{M}_o,$$

where $\{F_i\}$ are the Fock operators

$$(3.IV)\quad F_i = \frac{1}{2}\omega_i h + \sum_j (\alpha_{ij} J_j - \beta_{ij} K_j); \qquad i \in \mathcal{M}_o.$$

$\{\lambda_{ij}\}$ is the set of Lagrange multipliers, associated with the orthonormality conditions of the occupied MO set $\{|i\rangle\}$; h is the monoelectronic hamiltonian and $\{J_i\}$, $\{K_i\}$ are the Coulomb and exchange operators. The form of equations (2a.IV) and (2b.IV) fully depends on the structure of the exchange operators $\{K_i\}$ and characterizes monoconfigurational procedures. However, when one observes carefully the physical meaning of the $\{J_{ij}\}$ and $\{K_{ij}\}$ integrals, it is evident that the definition of the operators $\{K_j\}$, involved in the construction of the Fock operators is, in some manner, a computational artifact. So, a more suitable operator set may be defined as (see Section 3.II)

$$(4.IV)\quad V_{ij}(2) = \int \rho_{ij}(1) r_{12}^{-1}\, dr(1);\ i,\ j\ \varepsilon \mathcal{M}_0,$$
$$= \langle i(1)|r_{12}^{-1}|j(1)\rangle,$$

$\rho_{ij}(1)$ being a general charge distribution for electron 1, constructed from the MO's $|i\rangle$ and $|j\rangle$. Consequently, from this electrostatic point of view, both Coulomb and exchange integrals can be redefined as coulombic interaction energies between generalized continuous charge distributions, associated with electrons 1 and 2; that is:

$$(5a.IV)\quad J_{ij} = \iint \rho_{jj}(1)\rho_{ii}(2) r_{12}^{-1}\, dr(1)dr(2) = \langle i(2)|V_{jj}(1)|i(2)\rangle$$

and

$$(5b.IV)\quad K_{ij} = \iint \rho_{ij}(1)\rho_{ij}(2) r_{12}^{-1}\, dr(1)dr(2) = \langle i(2)|V_{ij}(1)|j(2)\rangle.$$

The operators (4.IV) contain the Coulomb operators as a particular case and represent in a more adequate manner than exchange operators the electrostatic repulsion between the electron pairs.
Now a variation of the functional (1.IV), taking into account the operators (4.IV) yields

$$(6.IV)\quad \delta E = \sum_i \omega_i \langle \delta i|h|i\rangle + \sum_i \sum_j 2\alpha_{ij}\langle \delta i|V_{jj}|i\rangle$$
$$- \sum_i \sum_j 2\beta_{ij}\langle \delta i|V_{ij}|j\rangle + \text{conjugate terms.}$$

After manipulating eq. (6.IV) and introducing the orthogonality constraints, one finally obtains the Euler equations

$$(7a.IV)\quad \sum_j F_{ij}|j\rangle = \sum_j \lambda_{ji}|j\rangle; \qquad i\ \varepsilon \mathcal{M}_0,$$

$$(7b.IV)\quad \lambda_{ji} = \lambda_{ij}^{*}; \qquad i,\ j\ \varepsilon \mathcal{M}_0,$$

where

$$(8.IV)\quad F_{ii} = \omega_i h + \sum_j \gamma_{ij} V_{jj},$$

$$(9.IV)\quad F_{ij} = -\ 2\gamma_{ij} V_{ij}; \qquad i \neq j,$$

with

(10.IV) $\gamma_{ij} = 2(2\alpha_{ij} - \delta_{ij}\beta_{ij})$,

δ_{ij} being the Kronecker delta symbol.
The new Euler equations (7.IV) have the general structure of the multiconfigurational problems, but due to their monoconfigurational background should give, when solved, the same results as obtained from the usual equations (2.IV).
The Fock operators, described with the aid of the coulombic operators $\{V_{ij}\}$ of equation (4.IV), hold a highly significant connection with the operators deduced from the energy variation in the Hartree method. Indeed, if $\{\omega_i\}$ represents the occupation numbers of the MO's, the Hartree operators $\{F_i^H\}$ related to the corresponding monoconfigurational state will be given by

(11.IV) $F_i^H = \omega_i h + \omega_i \sum_j \omega_j V_{jj} + 2(\omega_i - 1 - \frac{1}{2}\omega_i^2)\, V_{ii}$.

It is easy to check the fact that the operator form (11.IV) coincides with the diagonal elements (8.IV) of the $\{F_{ij}\}$ operators. In order to visualize this, it is sufficient to show that the set of constants $\{\gamma_{ij}\}$ can be re-defined as

(12.IV) $\gamma_{ij} = \omega_i\omega_j + 2\delta_{ij}(\omega_i - 1 - \frac{1}{2}\,\omega_i^2)$.

As a result of the previous discussion, the off-diagonal operators (9.IV) represent a contribution due to the wavefunction antisymmetrisation in the Hartree-Fock method. The possible interest of this scheme (besides the coherent definition of the coulombic operators and the structure of the new Fock operators, which sheds light on the connection between Hartree and Hartree-Fock methods) is the possibility to study, without interferences derived from a multiconfigurational wavefunction, a coupling operator to be used in order to solve the Euler equations (7.IV). In a previous section we have proposed the use of a unique coupling operator in order to solve a set of Euler equations like those of equation (7.IV). The construction of this coupling operator has been made in a manner that will generalize the coupling operator technique used with equations of type (2.IV).

2. Results

The computational framework discussed in the previous section has

Table 1.IV

Results for the Li isoelectronic sequence. Configuration $1s^2 2s^1 (^2S)$. Basis set {1s,2s} STO. The figures correspond to the optimal energies, upon exponent variation

	Li	Be^+	B^{2+}	C^{3+}
exponents	1s 2.691	3.69	4.69	5.68
	2s 0.640	1.09	1.52	1.96
E(1s) (au)	-4.919067	-10.248091	-17.595494	-26.940297
C_{11}	0.997153	0.996089	0.996040	0.996705
C_{21}	0.016656	0.016228	0.014337	0.010460
E(2s) (au)	-0.194881	-0.660238	-1.376657	-2.341961
C_{12}	-0.18142	-0.255744	-0.293700	-0.321302
C_{22}	1.013386	1.028268	1.038340	1.046882
energy (au)	-7.418484	-14.259375	-23.351029	-34.278885
virial	0.99985	0.99982	0.99941	0.99976
$\Delta = \lambda^*_{21} - \lambda_{12}$	8.7×10^{-7}	2.9×10^{-7}	5.7×10^{-7}	8.5×10^{-6}

Table 1.IV (continued)

Results for the Li isoelectronic sequence. Configuration $1s^2 2s^1 (^2S)$. Basis set {1s,2s} STO. The figures correspond to the optimal energies, upon exponent variation

	N^{4+}	O^{5+}	F^{6+}	Ne^{7+}
exponents	6.68	7.67	8.66	9.66
	2.38	2.80	3.23	3.65
E(1s)(au)	-38.297045	-51.658942	-67.012189	-84.369074
C_{11}	0.996705	0.997197	0.997602	0.997778
C_{21}	0.010460	0.008620	0.007171	0.006530
E(2s)(au)	-3.554732	-5.015476	-6.723624	-8.678152
C_{12}	-0.336309	-0.347478	-0.357874	-0.364549
C_{22}	1.051863	1.055968	1.059826	1.062268
energy(au)	-48.278885	-64.114054	-82.196433	-102.52595
virial	0.99938	1.00009	1.00022	0.99980
$\Delta = \lambda_{21}^* - \lambda_{12}$	1.2×10^{-6}	1.5×10^{-6}	2.8×10^{-6}	3.4×10^{-6}

permitted us to study the behaviour of the generalized coupling operators without problems. The ground states of He and Be as well as the Li isoelectronic sequence have been studied for this purpose.
He, Li and Be ground state wavefunctions have been reproduced from previous classical calculations within the present formalism, using 3 and 4 STO's basis sets. The total energies are identical, as can be expected, to the data in the literature. It should be noted, however, that although He and Li orbital energies and eigenvectors are found identical to those in the literature, Be solutions behave in a different manner. This can be explained by considering that Be is a problem with two closed shells and in this case there exists a unitary transformation which, leaving the electronic energy invariant, transforms the eigenvalues and vectors into the canonical orbitals.
Due to the structure of the operators (8.IV) and (9.IV), when He and Be are calculated, one can use the null gradient coupling operator part directly. In the Li atom case, due to the fact that conditions (7b.IV) should be fulfilled, one is compelled to also use the operator of hermiticity conditions.
We have consequently studied the behaviour of this operator in full with the aid of the Li isoelectronic series, optimizing the exponents of a minimal {1s, 2s} STO basis set. For a basis set of this kind and without the hermitean conditions part, the SCF process converges in the first iterative step for any initial MO pair, and the solution falls into a false extremum. In practice, and with the purpose of simplifying the procedure, all the constants $\{\sigma_{pq}\}$ in operator (16.II) have been chosen with the same absolute value σ. The value of σ which has given the best convergence pattern in any case has been $\sigma = 1$. Also, at the self-consistent stage the values of the differences $\Delta = \lambda^*_{21} - \lambda_{12}$, which measure the degree of hermiticity of the off-diagonal Lagrange multipliers, have been computed to be $\approx 10^{-6}$ for a convergence criterion of the eigenvector components with the value of 10^{-6}.
Table 1.IV collects the exponents, eigenvalues and vectors, total energy, virial coefficient, and Δ value for all the minima of the Li isoelectronic series.

3. Final Remarks

It has been shown that SCF theory can be easily reformulated in order to obtain a Euler's equation form with a multiconfigurational structure. This fact, obviously, does not simplify the computational framework but permits one:

(a) to define a more physically connected set of Fock operators;

(b) to work on the behaviour of a coupling operator technique, which transforms the computational problem into a unique pseudo secular equation.

In this manner the whole theoretical framework of Hartree-Fock's theory appears to be beautifully coherent.

V. Paired Excitation Multiconfigurational SCF

Although Multiconfigurational SCF theory can be studied within the general formalism of Chapter II, we will analyze essentially here the framework proposed during 1966-67, by Veillard and Clementi, based on early work from various sources and continued until now by a good deal of well known papers.

The name Paired Excitation (PE) will be adopted here in order to obtain, if possible, a consensus which may permit one to distinguish the present procedure from other MCSCF computational structures. By PEMCSCF theory we denote a MC procedure, involving any multiconfigurational state function of arbitrary spin multiplicity, where each term in the linear combination of the wavefunction is obtained by replacing a doubly occupied MO pair by a virtual pair. The prefix complete (CMCPESCF), when used, will mean a procedure taking into account all the doubly occupied and virtual orbitals.

Our main purpose now is to find a general algorithm in order to obtain easily the total energy expression, and the set of implicit Fock operators, in such a manner that a standard coupling operator formalism can be used. From a computational point of view it will be desirable to have a general procedure which, under some controlled parameters, can solve in the same fashion mono or multiconfigurational SCF problems. The PE framework has been chosen because, as we will show later, it satisfies these conditions.

Finally, the last point of interest has been the need to obtain a PE procedure capable of studying excited states: It should be mentioned, for instance, that the mainline of computational tasks until now has been carried out on a Closed Shell basis. The present formalism will take this last situation as a particular case within the general MCPE framework.

1. Closed Shell MCPESCF Theory

A discussion of the framework proposed by Veillard and Clementi and its description from a simpler point of view will be useful for generalization purposes. A closed shell configuration wavefunction Φ_k can be associated with an index set $S_k = \{i_1, i_2 \ldots , i_n\}$ of double occupied orbitals. If there are 2n electrons in the system, and m virtual orbitals available, the number of paired excitation (PE) functions which can be obtained is: $N_c = \binom{n+m}{n}$. To each of the N_c functions Φ_k there is an associated index set S_k. The structure of the closed shell PE hamiltonian can be obtained from the previous definition of the PE functions; applying the Slater rules to $\langle\Phi_K|\mathcal{H}|\Phi_L\rangle$ there appear to be three well defined cases:

a) Diagonal terms, $\#(S_K \cap S_L) = n$
b) Two different spinorbitals, if: $\#(S_K \cap S_L) = n - 1$
c) Null terms, if: $\#(S_K \cap S_L) \leqslant n - 2$,

where the symbol #(set) denotes the cardinal number of the set. So, from here it is easy to see the simplicity of the MCPESCF procedures, only two kinds of contributions to the MC energy will be present, namely:

(1a.V) $$\langle\Phi_K|\mathcal{H}|\Phi_K\rangle = 2 \sum_{i \varepsilon S_K} h_{ii} + \sum_{i,j \varepsilon S_K} (2\, J_{ij} - K_{ij})$$

and

(1b.V) $$\langle\Phi_K|\mathcal{H}|\Phi_L\rangle = \delta(n - 1 = \#(S_K \cap S_L))\; \delta(r \varepsilon S_K; s \varepsilon S_L; r,s \notin (S_K \cap S_L))\; K_{rs}.$$

In equation (1b.V) we use a <u>logical Kronecker's delta</u>, defined as $\delta(C_1;C_2;\ \ldots)$ subject to a set of conditions $\{C_i\}$; then $\delta(C_1;C_2;\ \ldots) = 1$ if and only if all C_i are simultaneously true, otherwise $\delta(C_1;C_2;\ \ldots) = 0$.

2. Complete MCPE Energy Expression

Let's suppose that a closed shell ground state wavefunction Ψ is built up with all the N_C configurations available in a given problem, then if

$$\Psi = \sum_K C_K \Phi_K, \text{ with, } \langle\Psi|\Psi\rangle = \sum_K C_K^2 = 1,$$

one can write

(2.V) $$E = \sum_K C_K^2 \langle\Phi_K|\mathcal{H}|\Phi_K\rangle + \sum_{\substack{K,L \\ (K \neq L)}} \delta(n - 1 = \#(S_K \cap S_L))\, C_K C_L \langle\Phi_K|\mathcal{H}|\Phi_L\rangle .$$

From here it is moderately easy to show that

$$E = \sum_{i \varepsilon \mathcal{P}} \omega_i h_{ii} + \sum_{i,j \varepsilon \mathcal{P}} (A_{ij} J_{ij} - B_{ij} K_{ij}),$$

where

(3.V) $$\omega_i = 2 \sum_K \delta(i \varepsilon S_K)\, C_K^2$$

(4.V) $$A_{ij} = 2 \sum_K \delta(i \varepsilon S_K;\ j \varepsilon S_K)\, C_K^2$$

(5.V) $$B_{ij} = \frac{1}{2} A_{ij} - 2 \sum_{\substack{K,L \\ (L>K)}} \delta(n - 1 = \#(S_K \cap S_L)) \times$$

$$\delta(i \varepsilon S_K;\ j \varepsilon S_L;\ i,j \notin (S_K \cap S_L))\, C_K C_L \ .$$

The sums in the energy run over all the orbitals entering in the set $\mathscr{P} = \bigcup_K S_K$. As one can easily see, the Fock operators will take a monoconfigurational form, with the state parameters substituted by the new sets $\{\omega_i\}$, $\{A_{ij}\}$ and $\{B_{ij}\}$; in this context the coupling operator procedure outlined in section 9.II will be entirely applicable in this case.

3. Two Electron Systems

As an application example of the framework studied in the previous section, a two electron problem will be studied here. In this case, each configuration will be associated with an index set of cardinal number unity and each configuration index will be made equal to a MO index. Then, if n MO's are used in the PEMCSCF procedure:

$$S_K = \{k\};\ (k = 1, n)\ ,$$

also the state parameters will be in this situation:

$$\omega_i = 2\, C_i^2$$

$$A_{ij} = 2\, \delta_{ij}\, C_i^2$$

$$B_{ij} = \delta_{ij}\, C_i^2 - (1 - \delta_{ij})\, C_i C_j$$

the energy will be given by

$$E = 2 \sum_i C_i^2 (h_{ii} + J_{ii}) - \sum_{i,j} B_{ij} K_{ij}$$

and the Fock operators are in this case:

$$F_i = \frac{1}{2}\,\omega_i\,(h + 2\,J_i) - \sum_j B_{ij}\,K_j;\quad (i = 1,\, n).$$

4. MCPESCF Theory with an Invariant Closed Shell

For the sake of completeness and in order to introduce into the computational scheme some closed shell orbitals, which do not play an active role in the PE system, one should define an index set $\mathscr{C}$ containing all the orbital indices not entering the PE movements, besides another set $\mathscr{P}$ which contains the remaining orbital indices used in the PE. It should be noted that the set $\mathscr{C}$ is a subset of all index sets $\{S_K\}$, associated to each PE wavefunction. Also:

$$\mathscr{C} \cup \mathscr{P} = \bigcup_K S_K.$$

The state parameters for all the indexes of $\mathscr{P}$ will have the forms studied previously: (2.V), (4.V) and (5.V).
On the other hand:

a) $\forall\, i \,\varepsilon\, \mathscr{C} : \omega_i = 2 \sum_K \delta(i\varepsilon S_K)\, C_K^2 = 2 \sum_K C_K^2 = 2$

because

$$\forall\, K:\quad i\varepsilon S_K$$

b) $\forall\, i \,\varepsilon\, \mathscr{C}$ and $\forall\, j \,\varepsilon\, \mathscr{C} : A_{ij} = 2 \sum_K \delta(i,j\varepsilon S_K)\, C_K^2 = 2 \sum_K C_K^2 = 2$

c) $\forall\, i \,\varepsilon\, \mathscr{C}$ and $\forall\, j \,\varepsilon\, \mathscr{P} : A_{ij} = 2 \sum_K \delta(j\varepsilon S_K)\, C_K^2 = \omega_j.$

Finally,

d) if $i,j \,\varepsilon\, \mathscr{C}$ or $i \,\varepsilon\, \mathscr{C}$, $j \,\varepsilon\, \mathscr{P} : B_{ij} = \frac{1}{2} A_{ij} = \frac{1}{2}\omega_j$,

because

$$i\varepsilon (S_K \cap S_L),\ \forall K,L.$$

Taking the above considerations into account, one can write:

1) $\forall\, i \,\varepsilon\, \mathscr{C} : F_i = F_o + \frac{1}{2} \sum_{j\varepsilon\mathscr{P}} \omega_j (2\,J_j - K_j)$

and

2) $\forall\, i \in \mathscr{P} : F_i = \frac{1}{2}\,\omega_i\, F_o + \sum_{j\in\mathscr{P}} (A_{ij}J_j - B_{ij}K_j)$

with the auxiliary operator F_o defined as

$$F_o = h + \sum_{j\in\mathscr{P}} (2\, J_j - K_j).$$

5. Open Shell PEMCSCF

Having developed the closed shell PEMC theory, the next task will be the construction of a general framework, valid for the widest possible range of state functions. The main idea underlying such a procedure should consist of an MC formalism, where the system's Fock operators remain diagonal, and consequently, where the coupling operator procedure outlined at the beginning can be easily applied.
It will be shown that this kind of algorithm can be studied and used, as in the closed shell case, under some restrictive conditions. In order to obtain this comfortable situation some previous definitions and a necessary theorem should be described. A paired excitation function (PEF) basis set is a set of state functions obtained by the replacement of coupled spinorbital pairs $\{\phi_d\bar{\phi}_d\}$, associated with the same spatial MO, by coupled virtual spinorbital pairs $\{\phi_v\bar{\phi}_v\}$ in a given initial state function. Let us write an initial state function, associated with an energy expression of monoconfigurational type as

$$\Phi_p = \sum_K A_K |\ldots\ \phi_p\bar{\phi}_p\ \ldots|_K$$

$\{|\ldots\ \phi_p\bar{\phi}_p\ \ldots|_K\}$ being a basis set of Slater Determinants (SD), with the following characteristics:

a) $\forall K$, the pair $\{\phi_p\bar{\phi}_p\}$ appears in the SD basis set in the same position.

b) Any two SD of the set differ by at least one spinorbital.

Also

$$\langle\Phi_p|\Phi_p\rangle = \sum_{K,L} A_K A_L \langle|\ldots\ \phi_p\bar{\phi}_p\ \ldots|_K \,||\ldots\ \phi_p\bar{\phi}_p\ \ldots|_L\rangle$$

$$= \sum_{K,L} A_K A_L\ \delta_{KL} = \sum_K A_K^2 = 1\ .$$

Then, without loss of generality any paired excitation function can be constructed from Φ_p as

$$\Phi_q = \sum_K A_K |\ldots\ \phi_q\bar{\phi}_q\ \ldots|_K\ ,$$

where the couple $\{\phi_p\bar{\phi}_p\}$ of Φ_p has been substituted by $\{\phi_q\bar{\phi}_q\}$. In this manner, the PEF basis set can be constructed from the set of all functions $\{\Phi_q\}$ associated with the primitive functions Φ_p, and it is not necessary to consider other types of paired excitations, because they do not interact with Φ_p.

Now, we can easily show the validity of the following theorem.

Theorem: (Structure of the PE hamiltonian).

The off-diagonal hamiltonian matrix elements with respect of the PEF basis set have the following algorithm

$$\langle\Phi_p|\mathscr{H}|\Phi_q\rangle = (pq|pq); \quad (p \neq q). \tag{6.V}$$

Proof:

$$\langle\Phi_p|\mathscr{H}|\Phi_q\rangle = \sum_{K,L} A_K A_L \langle|\ldots \phi_p\bar{\phi}_p \ldots|_K \mathscr{H} |\ldots \phi_q\bar{\phi}_q \ldots|_L\rangle$$

$$= \sum_K A_K^2 \langle|\ldots \phi_p\bar{\phi}_p \ldots|_K \mathscr{H} |\ldots \phi_q\bar{\phi}_q \ldots|_K\rangle.$$

This follows since between any pair of SD with $K \neq L$ there exist, by construction, more than three differences. Then, between two SD with $K = L$ there are only two differences, and a maximum coincidence. Slater's rules give (see Appendix B)

$$\langle\Phi_p|\mathscr{H}|\Phi_q\rangle = \sum_K A_K^2 \, (pq|pq),$$

and using the normalization condition, one obtains the proposed result.■

At the same time, the hamiltonian diagonal elements can be written as

$$\begin{aligned}\langle\Phi_t|\mathscr{H}|\Phi_t\rangle = {} & 2 \sum_{p\in\mathscr{C}} h_{pp} + \sum_{p\in\mathscr{C}} \sum_{q\in\mathscr{C}} (2\, J_{pq} - K_{pq}) + \sum_{k\in\mathscr{O}} v_k h_{kk} \\ & + \sum_{k\in\mathscr{O}} \sum_{\ell\in\mathscr{O}} (\alpha_{k\ell} J_{k\ell} - \beta_{k\ell} K_{k\ell}) + 2 \sum_{r\in\mathscr{P}_t} h_{rr} \\ & + \sum_{p\in\mathscr{C}} \sum_{k\in\mathscr{O}} v_k (2\, J_{pk} - K_{pk}) \\ & + \sum_{r\in\mathscr{P}_t} \sum_{s\in\mathscr{P}_t} (2\, J_{rs} - K_{rs}) + 2 \sum_{p\in\mathscr{C}} \sum_{r\in\mathscr{P}_t} (2\, J_{pr} - K_{pr}) \\ & + \sum_{k\in\mathscr{O}} v_k \sum_{r\in\mathscr{P}_t} (2\, J_{kr} - K_{kr})\end{aligned} \tag{7.V}$$

where $\mathscr{C}$ is the index set of doubly occupied orbitals not considered in the PE; $\mathscr{O}$ is the set of indexes of the open shell orbitals, which cannot enter the PE process and should remain in the same relative position throughout the construction of PEF; and $\mathscr{P}_t$ is the set of pair replacement orbital indexes, entering the state function Φ_t. $\{v_k\}$, $\{\alpha_{k\ell}\}$ and $\{\beta_{k\ell}\}$ are the state parameters associated with the PEF. If the PEMC function is constructed by means of a similar expression as in the closed shell case, with S_K containing the set of doubly occupied orbitals in a given configuration, then the energy can be written with an expression alike to equation (2.V). Taking into account the previous expressions (6.V) and (7.V), one can write:

$$E = E(\mathscr{C}, \mathscr{O}) + \sum_{r\varepsilon\mathscr{P}} \sum_{s\varepsilon\mathscr{P}} (A_{rs}J_{rs} - B_{rs}K_{rs})$$

$$+ \sum_{r\varepsilon\mathscr{P}} \omega_r [h_{rr} + \sum_{p\varepsilon\mathscr{C}} (2\,J_{pr} - K_{pr}) + \frac{1}{2} \sum_{k\varepsilon\mathscr{O}} v_k (2\,J_{kr} - K_{kr})]$$

where $\mathscr{C}$, $\mathscr{O}$ have the meaning discussed previously; $\mathscr{P}$ is the set of all MO indexes entering the PEF; $E(\mathscr{C}, \mathscr{O})$ is the part of expression (7.V) where only sums over $\mathscr{C}$, $\mathscr{O}$ or both index sets are performed. That is, the first four terms in the equation; and $\{\omega_r\}$, $\{A_{rs}\}$ and $\{B_{rs}\}$ are state parameters, in this case:

a) $\omega_r = 2 \sum_K \delta(r\varepsilon S_K)\, C_K^2$

b) $A_{rs} = 2 \sum_K \delta(r,s\varepsilon S_K)\, C_K^2$

c) $$B_{rs} = \sum_K \delta(r,s\varepsilon S_K)\, C_K^2 - \sum_{\substack{K,L \\ (K\neq L)}} \delta(n - 1 = \#(S_K \cap S_L))$$

$$\delta(r\varepsilon S_K;\ s\varepsilon S_L;\ r,s\notin(S_K \cap S_L))\, C_K C_L.$$

From this, the Fock operators are written as:

1) $\forall\ p\ \varepsilon\mathscr{C}:\quad F_p = F_o + \frac{1}{2}(G_{\mathscr{O}} + G_{\mathscr{P}}) = F_{\mathscr{C}}$

2) $\forall\ k\ \varepsilon\mathscr{O}:\quad F_k = \frac{1}{2} v_k (F_o + \frac{1}{2} G_{\mathscr{P}}) - \sum_{i\varepsilon\mathscr{O}} (\alpha_{k\ell}J_\ell - \beta_{k\ell}K_\ell)$

3) $\forall\ r\ \varepsilon\mathscr{P}:\quad F_r = \frac{1}{2} \omega_r (F_o + \frac{1}{2} G_{\mathscr{O}}) + \sum_{s\varepsilon\mathscr{P}} (A_{rs}J_s - B_{rs}K_s)$,

with the auxiliary operators F_o, $G_{\mathscr{O}}$ and $G_{\mathscr{P}}$ defined as

a) $F_o = h + \sum (2\,J_q - K_q)$

b) $G_{\mathscr{O}} = \sum_{k\varepsilon\mathscr{O}} v_k\ (2\,J_k - K_k)$

c) $G_{\mathscr{P}} = \sum_{r\varepsilon\mathscr{P}} \omega_r (2\, J_r - K_r)$.

The total energy can be written

$$E = \sum_{p\varepsilon\mathscr{C}} \langle p|F_{\mathscr{C}} + h|p\rangle + \sum_{k\varepsilon\mathscr{O}} \langle k|F_k + \frac{1}{2}\, v_k\, h\,|k\rangle$$

$$+ \sum_{r\varepsilon\mathscr{P}} \langle r|F_r + \frac{1}{2}\, \omega_r\, h\,|r\rangle .$$

In fact, using the obtained results, one can also consider the energy expressed in a more compact form as

$$E = \sum_t N_t h_{tt} + \sum_{t,u} (A_{tu} J_{tu} - B_{tu} K_{tu})$$

with the state parameters defined by the scheme

		$\mathscr{C}$	$\mathscr{O}$	$\mathscr{P}$
N		2	v	ω
A	$\mathscr{C}$	2	v	ω
	$\mathscr{O}$		α	$\frac{1}{2}\, v\cdot\omega$
	$\mathscr{P}$			A
B	$\mathscr{C}$	1	$\frac{1}{2} v$	$\frac{1}{2}\, \omega$
	$\mathscr{O}$		β	$\frac{1}{4}\, v\cdot\omega$
	$\mathscr{P}$			B

Then, the Fock operators can be constructed in a compact form as:

$$F_t = \frac{1}{2}\, N_t\, h + \sum_u (A_{tu} J_u - B_{tu} K_u).$$

Finally, the energy can also be written, by means of the new Fock operators, as

$$E = \sum_t \langle t|F_t + \frac{1}{2}\, N_t h|t\rangle.$$

6. Special Cases

Some special cases can be derived from this general formalism:

a) Monoconfigurational Case: the set $\mathscr{P} = \emptyset$ and only the operators associated with the sets $\mathscr{C}$ and $\mathscr{O}$ should be considered.

b) <u>Multiconfigurational Closed Shell Case</u>: In this situation $\mathcal{O} = \emptyset$, and one finds again the operators discussed in the previous analysis of Clementi and Veillard formalism.

c) <u>Complete Multiconfigurational Open Shell Case</u>: Here, the set $\mathcal{C} = \emptyset$. The operator $F_{\mathcal{C}} = h$, and one should only consider the operators associated with $\mathcal{O}$ and $\mathcal{P}$ sets.

d) <u>Complete Multiconfigurational Closed Shell Case</u>: Then simultaneously $\mathcal{C}$ and $\mathcal{O} = \emptyset$, and only the operators of the $\mathcal{P}$ set should be taken into account.

VI. SCF Perturbation Theory

1. Perturbation Theory

1.1. General Scheme

Consider the secular equation of an unperturbed hermitean operator R_o, written as

$$R_o|o;i> = \varepsilon_{o;i}|o;i> \tag{1.VI}$$

and construct a perturbation operator V as a power series of some parameter λ:

$$V = \sum_{k=1}^{\infty} \lambda^k V_k , \tag{2.VI}$$

using $\{V_k\}$ as a set of hermitean operators. A perturbed operator R can be introduced through the relation

$$R = R_o + V, \tag{3.VI}$$

where R satisfies the secular equation

$$R|i> = \varepsilon_i|i>. \tag{4.VI}$$

An approximate set of solutions can be obtained through the expansions

$$|i> = \sum_{k=o}^{\infty} \lambda^k|k;i> \tag{5.VI}$$

and

$$\varepsilon_i = \sum_{k=o}^{\infty} \lambda^k \varepsilon_{k;i} . \tag{6.VI}$$

Substituting (2.VI), (5.VI) and (6.VI) into (4.VI), and equating terms of equal power in λ, an infinite set of equations is found;

$$\begin{aligned} R_o|n;i> + (1 - \delta_{no}) \sum_{k=1}^{n} V_k|n-k;i> \\ = \sum_{k=o}^{n} \varepsilon_{k;i}|n-k;i>;\ n = 0, 1, \ldots . \end{aligned} \tag{7.VI}$$

Equation (7.VI) can be transformed into a more convenient form, by defining the hermitean operators

$$\Omega_{k;i} = V_k - \varepsilon_{k;i}\, I, \tag{8.VI}$$

where I is the unit operator. Then, using (8.VI) in (7.VI) the nth order equation, and dropping the index i in $\Omega_{k,i}$,

$$-\,\Omega_o|n;i\rangle = \Omega_n|o;i\rangle - (1-\delta_{n1})\sum_{k=1}^{n-1}\Omega_k|n-k;i\rangle, \tag{9.VI}$$

is found, which will be used to obtain the nth order corrections $|n;i\rangle$ and $\varepsilon_{n;i}$ to the eigenvalues and eigenvectors of R_o.

1.2. Orthonormalization Conditions

It is interesting to note that, because R_o is hermitean, the eigenvector set $\{|o;i\rangle\}$ can be considered orthonormalized and complete. Therefore, the conditions

$$\langle i;o|o;j\rangle = \delta_{ij}, \tag{10.VI}$$

always hold. The eigenvectors of R_o can be used as a basis set in order to construct the eigenvector corrections,

$$|k;i\rangle = \sum_{j\neq i} a_{k;ij}|o;j\rangle \quad (k>o), \tag{11.VI}$$

$\{a_{k;ij}\}$ being a set of coefficients to be determined. Using (11.VI) and (10.VI):

$$\begin{aligned}\langle i;o|k;i\rangle &= \sum_{j\neq i} a_{k;ij}\langle i;o|o;j\rangle\\ &= \sum_{j\neq i} a_{k;ij}\,\delta_{ij}\\ &= 0 \qquad (k>o),\end{aligned}$$

and, in general, one can write

$$\langle i;o|k;j\rangle = \delta_{ko}\,\delta_{ij} + (1-\delta_{ij})(1-\delta_{ko})\,a_{k;ij}. \tag{12.VI}$$

Also, if the orthonormality conditions of the perturbed operator eigenvectors are taken into account

$$<i|j> = \delta_{ij}, \tag{13.VI}$$

then using (5.VI),

$$\sum_{k=o}^{n} <i;k|n-k;j> = 0 \qquad (n>o), \tag{14.VI}$$

so from equation (12.VI) using (11.VI):

$$\sum_{k=o}^{n} \sum_{p\neq i,j} a^{*}_{k;ip}\, a_{n-k;jp} = 0 \ ; \quad (n>o). \tag{15.VI}$$

1.3. <u>Eigenvalue Corrections</u>

Equation (9.VI) multiplied on the left by $<i;o|$ gives

$$\begin{aligned} - <i;o|\Omega_o|n;i> &= <i;o|\Omega_n|o;i> \\ &+ (1-\delta_{n1}) \sum_{k=1}^{n-1} <i;o|\Omega_k|n-k;i> \ , \end{aligned} \tag{16.VI}$$

but if one looks at each term of (16.VI):

$$<i;o|\Omega_o = <i;o|R_o - \varepsilon_{o;i}<i;o| \ = <0|$$

and

$$<i;o|\Omega_n|o;i> = <i;o|V_n|o;i> - \varepsilon_{n;i} \ ,$$

then (16.VI) can be written as

$$\varepsilon_{n;i} = <i;o|V_n|o;i> + (1-\delta_{n1}) \sum_{k=1}^{n-1} <i;o|\Omega_k|n-k;i> \ , \tag{17.VI}$$

which gives the eigenvalue corrections up to any order.

1.4. <u>Wigner's Theorem</u>

Wigner's Theorem states that, when the nth eigenvector corrections are known, then the 2n and 2n+1 eigenvalue corrections can be computed without the need of higher order vector expressions.
This follows from the hermitean nature of the operators $\{\Omega_{k;i}\}$, using

recursively the nth order equation (17.VI), for $n > 3$. The case $n = 3$ will be shown here. From equation (17.VI):

$$\varepsilon_{3;i} = \langle i;o|V_3|o;i\rangle + \langle i;o|\Omega_1|2;i\rangle + \langle i;o|\Omega_2|1;i\rangle \quad , \tag{18.VI}$$

also equation (9.VI) gives

$$- \Omega_o|2;i\rangle = \Omega_2|o;i\rangle + \Omega_1|1;i\rangle$$

and

$$- \Omega_o|1;i\rangle = \Omega_1|o;i\rangle.$$

From these equalities one can deduce that

$$\langle i;o|\Omega_1|2;1\rangle = - \langle i;1|\Omega_o|2;i\rangle$$

$$\langle i;1|\Omega_1|1;i\rangle + \langle i;1|\Omega_2|o;i\rangle \ ,$$

and substituting this result into (18.VI) gives

$$\varepsilon_{3;i} = \langle i;o|V_3|o;i\rangle + \langle i;1|\Omega_1|1;i\rangle + \langle i;1|\Omega_2|o;i\rangle + \langle i;o|\Omega_2|1;i\rangle \ , \tag{19.VI}$$

where only the first order correction $|1;i\rangle$ is needed to obtain $\varepsilon_{3,i}$. The procedure can be applied up to any order, and the generalized form for $n > 3$ and $2 < 2\,k < n - 1$ is

$$\varepsilon_{n;i} = \langle i;o|V_n|o;i\rangle + \sum_{p=1}^{n-k-1} \langle i;o|\Omega_{n-p;i}|p;i\rangle + \sum_{p=1}^{k} \sum_{q=p}^{n-(k-p+1)} \langle i;p|\Omega_{n-q;i}|q-p;i\rangle \ . \tag{20.VI}$$

1.5. Eigenvector Corrections

Suppose a non-degenerate spectrum in R_o exists, then the eigenvector corrections can be found by multiplying equation (9.VI) on the left by $\langle j;o|$. One obtains the expression

$$- \langle j;o|\Omega_{o;i}|n;i\rangle = \langle j;o|\Omega_{n;i}|o;i\rangle$$

$$+ (1-\delta_{n1}) \sum_{k=1}^{n-1} <j;o|\Omega_{k;i}|n-k;i> , \qquad (21.VI)$$

which can be simplified by taking into account

$$<j;o|\Omega_{o;i}|n;i> = <j;o|R_o - \varepsilon_{o;i}I|n;i>$$

$$= (\varepsilon_{o;j} - \varepsilon_{o;i})\, a_{n;ij},$$

so that finally one obtains the relation

$$a_{n;ij} = (\varepsilon_{o;i} - \varepsilon_{o;j})^{-1} [<j;o|V_n|o;i> + (1 - \delta_{n1}) \sum_{k=1}^{n-1} \{<j;o|V_k|n-k;i> - \varepsilon_{k;i} a_{n-k;ij}\}] . \qquad (22.VI)$$

1.6. Alternative Formalism

Another procedure can be used which utilizes the definition of the operators

$$T_i = \sum_{j\neq i} (\varepsilon_{o;i} - \varepsilon_{o;j})^{-1} |o;j><j;o| . \qquad (23.VI)$$

Thus, another operator set $\{Q_{n;i}\}$ shall be defined in order to fulfill the equations

$$|n;i> = T_i Q_{n;i}|o;i> . \qquad (24.VI)$$

It is easy to obtain

$$T_i |o;i> = |0> ,$$

so to $Q_{n;i}$ can be added any operator of the form αI, α being an arbitrary scalar, without modifying equation (24.VI). In order to overcome this arbitrary situation, it will be supposed that

$$<i;o|Q_{n;i}|o;i> = 0 . \qquad (25.VI)$$

Now, given an arbitrary vector $|a>$, we have

$$- \Omega_{o;i}T_i|a> = - (R_o - \varepsilon_{o;i}I) \sum_{j\neq i} (\varepsilon_{o;i} - \varepsilon_{o;j})^{-1} |o;j><j;o|a>$$

$$= \sum_{j\neq i} (\varepsilon_{o;i} - \varepsilon_{o;j})^{-1} (\varepsilon_{o;i} - \varepsilon_{o;j})|o;j\rangle\langle j;o|a\rangle$$

$$= \sum_{j\neq i} |o;j\rangle\langle j;o|a\rangle.$$

But, the set $\{|i;o\rangle\}$ being complete enables one to write

$$I = \sum_{j} |j;o\rangle\langle o;j| \ ,$$

and using this, the previous equation is rewritten as

$$- \Omega_{o;i}T_i|a\rangle = |a\rangle - |o;i\rangle\langle i;o|a\rangle \ . \qquad (26.VI)$$

Finally using (23.VI), (25.VI) and (26.VI) into (9.VI), one has:

$$- \Omega_{o;i}T_iQ_{n;i}|o;i\rangle = \Omega_{n;i}|o;i\rangle - (1 - \delta_{n1}) \sum_{k=1}^{n-1} \Omega_{k;i}T_iQ_{n-k;i}|o;i\rangle \ ,$$

and in this manner $Q_{n;i}$ can be defined up to any order by

$$Q_{n;i} = \Omega_{n;i} + (1 - \delta_{n1}) \sum_{k=1}^{n-1} \Omega_{k;i}T_iQ_{n-k;i} \ . \qquad (27.VI)$$

Using the set $\{Q_{n;i}\}$, the eigenvalue corrections can be obtained through the expression

$$\varepsilon_{n;i} = \langle i;o|U_{n;i}|o;i\rangle \ ,$$

where

$$U_{n;i} = Q_{n;i} + \varepsilon_{n;i}\, I$$

$$= V_n - (1 - \delta_{n1}) \sum_{k=1}^{n-1} \Omega_{k;i}T_i\, Q_{n-k;i} \ . \qquad (28.VI)$$

2. Open Shell SCF Theory

2.1. Energy

The unperturbed SCF energy can be written as

$$E_o = \sum_i \langle i;o|\omega_i h + \sum_j (\alpha_{ij} J_{oo;j} - \beta_{ij} K_{oo;j})|o;i\rangle \,, \tag{29.VI}$$

where in order to have a more convenient expression the repulsion term can be written with the aid of the operators $\{X_{oo;i}\}$, defined by means of the unperturbed Coulomb $\{J_{oo;j}\}$ and exchange $\{K_{oo;j}\}$ operators, through

$$X_{oo;j} = \sum_j (\alpha_{ij} J_{oo;j} - \beta_{ij} K_{oo;j}) \,. \tag{30.VI}$$

Using this definition, (29.VI) can be rewritten as

$$E_o = \sum_i \langle i;o|\omega_i h + X_{oo;i}|o;i\rangle \,. \tag{31.VI}$$

If the electronic hamiltonian of the system is perturbed by a monoelectronic operator u, then the perturbed energy can be written as

$$E = \sum_i \langle i;o|\omega_i (h + u) + X_i|o;i\rangle \,, \tag{32.VI}$$

where

$$X_i = \sum_j (\alpha_{ij} J_j - \beta_{ij} K_j) \,. \tag{33.VI}$$

The perturbed energy can be written as an infinite power series

$$E = \sum_{k=o}^{\infty} \lambda^k E_k \,, \tag{34.VI}$$

and each term $\{E_k\}$ in (34.VI) can be easily expressed as

$$E_k = \sum_i \sum_{p=o}^{k} \sum_{q=o}^{k-p} \langle i;p|Y_{k-(p+q);i}|q;i\rangle \,, \tag{35.VI}$$

with the operators $\{Y_\ell\}$ defined as

$$Y_{\ell;i} = \delta_{\ell o}\, \omega_i h + \delta_{\ell 1}\, \omega_i u + \sum_{r=o}^{\ell} X_{r,\ell-r;i} \,, \tag{36.VI}$$

and $X_{p,q;i}$ defined through the expression

$$X_{p,q;i} = \sum_j (\alpha_{ij} J_{p,q;j} - \beta_{ij} K_{p,q;j}) \,, \tag{37.VI}$$

$\{J_{p,q;j}\}$ and $\{K_{p,q;j}\}$ being Coulomb and exchange operators respectively, calculated by means of the pth and q^{th} corrections to the jth MO.

2.2. Eigenvector Corrections

From the previous discussion, it necessarily follows that the key to obtaining the energy corrections is the MO corrections. They can be obtained through the solution of the pseudosecular equation of the perturbed coupling operator R:

$$R|i\rangle = \varepsilon_{ii}|i\rangle \quad . \tag{38.VI}$$

The Coupling operator is defined through the expression

$$R = \sum_i \Pi_i F_i \Pi_i + \sum_i \sum_j \Theta_{ij}\{P_j(F_i - F_j)P_i\} \tag{39.VI}$$

with

$$F_i = \frac{1}{2}\omega_i(h + u) + X_i,$$

$$P_i = |i\rangle\langle i|$$

and

$$\Pi_i = P_i + \sum_v |v\rangle\langle v| \ ,$$

the last sum being carried over the virtual orbital indexes.
The Coupling operator can be developed as a series

$$R = \sum_{k=o}^{\infty} \lambda^k R_k \ , \tag{40.VI}$$

by defining the auxiliary operators

$$A_{pqr} = \sum_i \Pi_{p;i} F_{q;i} \Pi_{r;i}$$

$$+ \sum_i \sum_j \Theta_{ij} P_{p;j} (F_{q;i} - F_{q;j}) P_{r;i}, \tag{41.VI}$$

so that

$$R_k = \sum_{p=o}^{k} \sum_{q=o}^{k-p} A_{p,q,k-(p+q)} \; . \tag{42.VI}$$

Then, the formalism developed in the previous section can be applied without further rearrangements.

In order to complete the construction of the set $\{A_{pqr}\}$, the operators $\{\Pi_{p;i}\}$, $\{F_{p;i}\}$ and $\{P_{p;i}\}$ are needed. The structure of these operators can be deduced without effort from the corresponding perturbed definitions, in this manner we have

$$P_{p;i} = \sum_{r=o}^{p} |r;i\rangle\langle i;p-r| \; , \tag{43.VI}$$

$$\Pi_{p;i} = P_{p;i} + \sum_{v} \sum_{r=o}^{p} |r;v\rangle\langle v;p-r| \tag{44.VI}$$

and

$$F_{p;i} = \frac{1}{2}\omega_i(\delta_{po}h + \delta_{p1}u) + X_{p;i} . \tag{45.VI}$$

As an example, the first order terms will be

a) $F_{1;i} = \frac{1}{2}\,\omega_i u + X_{1;i}$

b) $P_{1;i} = |o;i\rangle\langle i;1| + |1;i\rangle\langle o;i|$

c) $\Pi_{1;i} = P_{1;i} + \sum_{v}\{|o;v\rangle\langle v;1| + |1;v\rangle\langle v;o|\}$,

consequently

$$R_1 = \sum_{i}\{\Pi_{1;i}F_{o;i}\Pi_{o;i} + \Pi_{o;i}F_{1;i}\Pi_{o;i} + \Pi_{o;i}F_{o;i}\Pi_{1;i}\}$$

$$+ \sum_{i}\sum_{j} \Theta_{ij}\{P_{1;i}(F_{o;j} - F_{o;i})P_{o;j} + P_{o;i}(F_{1;j} - F_{1;i})P_{o;j}$$

$$+ P_{o;i}(F_{o;j} - F_{o;i})P_{1;j}\} \; .$$

2.3. Procedure

The main problem at each correction step lies in the dependence of the coupling operator perturbative terms on the MO corrections.

Due to this dependency the perturbation sequence shall be obtained at each level iteratively. For example, the nth order correction to the energy will be obtained through the following algorithm:

1) Use n - 1 order vectors to obtain $\sim R_n$, neglecting the terms Π_n, P_n and F_n.
2) Use of $\sim R_n$ within the perturbation formulae to obtain the nth order corrections on the eigenvectors $\{|n;i\rangle\}$.
3) $\{|n;i\rangle\}$ can be then used to construct Π_n, P_n and F_n as well as a new and more correct R_n.
4) Steps 2) and 3) can be repeated until the nth order energy correction becomes stable.

2.4. Closed Shell SCF Perturbation Theory

Closed Shell perturbation theory arises as a trivial simplification of the above structure. In this case we have

a) $F_k = F \; , \forall k$

with

$$F = h + u + X_c,$$

and

$$X_c = \sum_j (2 J_j - K_j).$$

b) $R = F$

so

$$R_k = \delta_{ko} h + \delta_{k1} u + \sum_{p=o}^{k} X_{c;p,k-p}$$

and

$$X_{c;p,k-p} = \sum_j (2 J_{p,k-p;j} - K_{p,k-p;j}).$$

3. Interaction of Two Molecules as an Application Example

3.1. General Background

Consider two electronic systems, A and B, separated and interacting in such a manner that the total wavefunction of the interacting supersystem can be written as a product

$$\Psi \sim \Psi_A \Psi_B ,$$

Ψ_A and Ψ_B being the wavefunctions of the respective isolated systems. Let's write the total electronic hamiltonian of the supersystem as

$$\mathcal{H}_{AB} = \sum_i \left(-\frac{1}{2} \nabla_i^2 - \sum_a \frac{Z_a}{r_{ai}} - \sum_b \frac{Z_b}{r_{bi}} + \sum_{j<i} \frac{1}{r_{ij}}\right),$$

where the indexes i,j are attached to the electrons of the supersystem and a, b are labelling the atoms of A and B, respectively.
The electronic energy of the whole system will be

$$E = \langle \Psi | \mathcal{H}_{AB} | \Psi \rangle ,$$

and the following integrals shall be calculated

a) $$\int \Psi_A^* \Psi_B^* \left(-\frac{1}{2} \nabla_i^2\right) \Psi_A \Psi_B \, dV_i = K ,$$

which if $i \in A$: $K = \langle \Psi_A | -\frac{1}{2} \nabla^2 | \Psi_A \rangle = K_A$

and if $i \in B$, $K = K_B$.

b) $$-\sum_a \int \Psi_A^* \Psi_B^* \frac{Z_a}{r_{ai}} \Psi_A \Psi_B dV_i - \sum_b \int \Psi_A^* \Psi_B^* \frac{Z_b}{r_{bi}} \Psi_A \Psi_B \, dV_i = V$$

then, in this case, we have

if $i \in A$: $$V_A = -\sum_a \langle \Psi_A | \frac{Z_a}{r_a} | \Psi_A \rangle - \sum_b \langle \Psi_A | \frac{Z_b}{r_b} | \Psi_A \rangle$$

$$= (AA:A) + (AA:B)$$

and if $i \in B$: $V_B = (BB:B) + (BB:A)$.

c) $$\sum_i \sum_{j<i} \int \Psi_A^* \Psi_B^* \left(\frac{1}{r_{ij}}\right) \Psi_A \Psi_B \, dV_i \, dV_j = L$$

In order to obtain an adequate expression for L, one can use the relation

$$\sum_i \sum_{j<i} \frac{1}{r_{ij}} = \frac{1}{2} \sum_i \sum_{j \neq i} \frac{1}{r_{ij}} ,$$

so that $L = \frac{1}{2} \{ \sum_{i \varepsilon A} \sum_{\substack{j \varepsilon A \\ j \neq i}} \langle \Psi_A | \frac{1}{r_{ij}} | \Psi_A \rangle$

$$+ \sum_{i \varepsilon B} \sum_{\substack{j \varepsilon B \\ j \neq i}} \langle \Psi_B | \frac{1}{r_{ij}} | \Psi_B \rangle \}$$

$$+ \sum_{i \varepsilon A} \sum_{j \varepsilon B} \langle \Psi_A \Psi_B | \frac{1}{r_{ij}} | \Psi_A \Psi_B \rangle \} .$$

From the structure of these three integrals, it can be seen that

$$E = E_A + E_B + E_{AB} ,$$

and with I = A, B, the terms E_A, E_B are given by

$$E_I = \sum_{p \varepsilon I} \langle p;I | \omega_p^I \, h^I + \sum_{q \varepsilon I} (\alpha_{pq}^{II} \, J_q^I - \beta_{pq}^{II} \, K_q^I) | I;p \rangle ,$$

and coincide with the isolated energies of A and B calculated with the supersystem MO's.
On the other hand

$$E_{AB} = (AA:B) + (BB:A) + \sum_{i \varepsilon A} \sum_{j \varepsilon B} \langle \Psi_A \, \Psi_B | \frac{1}{r_{ij}} | \Psi_A \, \Psi_B \rangle$$

can be associated with the interaction energy of the systems.
The necessary integrals can be evaluated in terms of the MO's of $A\{|A;p\rangle\}$ and $B\{|B;p\rangle\}$ as follows:

a) $$(AA:B) = - \sum_b \sum_{p \varepsilon A} \omega_p^A \langle p;A | \frac{Z_b}{r_b} | A;p \rangle$$

and an equivalent expression for (BB:A) can be also found.

b) Defining the Coulomb operator as

$$\sum_{i\varepsilon A} \int \Psi_A^*(i) \frac{1}{r_{ij}} \Psi_A(i)\, dV_i = \sum_{p\varepsilon A} \omega_p^A \int \phi_p^*(1) \frac{1}{r_{1j}} \phi_p(1)\, dV_1$$

$$= \sum_{p\varepsilon A} \omega_p^A J_p(j)$$

it follows that

$$\sum_{i\varepsilon A} \sum_{i\varepsilon B} \langle \Psi_A \Psi_B | \frac{1}{r_{ij}} | \Psi_A \Psi_B \rangle$$

$$= \sum_{q\varepsilon B} \omega_q^B \langle q;B| \sum_{p\varepsilon A} \omega_p^A J_p^A |B;q\rangle$$

$$= \sum_{p\varepsilon A} \omega_p^A \langle p;A| \sum_{q\varepsilon B} \omega_q^B J_q^B |A;p\rangle$$

$$= \sum_{p\varepsilon A} \sum_{q\varepsilon B} \omega_p^A \omega_q^B (pp|qq)$$

and the interaction energy can then be written as

$$E_{AB} = - \sum_b \sum_{p\varepsilon A} \omega_p^A \langle p;A| \frac{Z_b}{r_b} |A;p\rangle - \sum_a \sum_{q\varepsilon B} \omega_q^B \langle q;B| \frac{Z_a}{r_a} |B;q\rangle$$

$$+ \sum_{p\varepsilon A} \sum_{q\varepsilon B} \omega_p^A \omega_q^B (pp|qq) \; .$$

3.2. Variational Equations

The variation of E, the total energy of the supersystem, will be equivalent to varying each term separately. That is

$$\delta E = \delta E_A + \delta E_B + \delta E_{AB} \; ,$$

including the orthonormality conditions in the augmented functional, and using

$$V_I = - \sum_{c\varepsilon I} \frac{Z_c}{r_c} \; , \qquad I = A, B$$

then

$$h_I = - \frac{1}{2} \nabla^2 + V_I \; , \qquad I = A, B$$

and one can construct the Fock operators

$$p\varepsilon I : F_{I;p} = \frac{1}{2}\,\omega_p^I (h_I + V_J) + X_{I;p} + \frac{1}{2}\,\omega_p^I \sum_{q\varepsilon J} \omega_q^J\, J_{J;q}$$

where I = A, J = B or I = B, J = A and

$$X_p^I = \sum_{q\varepsilon I} (\alpha_{pq}^{II} J_{I;q} - \beta_{pq}^{II} K_{I;q}).$$

The Fock operators previously defined can be separated into two parts

a) $$F_{I;p}^{o} = \frac{1}{2}\,\omega_p^I\, h + X_{I;p}$$

b) $$U_J = \{V_J + \sum_{q\varepsilon J} \omega_q^J\, J_{J;q}\} ,$$

this last operator being a monoelectronic operator which accounts for the attraction of the electrons of I by the nucleus of J, (V_J), as well as the repulsion of the electronic cloud of I by the electronic cloud of J, $(\sum_{q\varepsilon J} \omega_q^J J_{J;p})$. The whole operator can then be written as

$$F_{I;p} = F_{I;p}^{o} + \frac{1}{2}\,\omega_p^I\, U_J .$$

One can see, at this point, that the whole problem has been transformed into a perturbation of the system I Fock operators by a set of monoelectronic operators $\frac{1}{2}\,\omega_p^I\, U_J$. The operator U_J, if one remembers the SCF perturbative formalism, acts as the monoelectronic perturbation which has been used in the preceding section.

3.3. Perturbational Scheme

The previous discussion shows that in the case where one wants to use a perturbational scheme each subsystem should be studied in turn. That is, a) Solve the SCF problem for each system.

b) Use the vectors of system J to construct U_J, and use this operator as a perturbation on system I.

c) Solve the perturbation equations on system I.

d) Use the vectors of I to construct U_I.

e) Solve the perturbation equations on system J.

3.4. Nature of the Interaction Energy

The nth order corrections to the total energy will have the following schematic structure:

$$E_n = E_{n,A} + E_{n,B} + E_{n,AB} .$$

As the interaction energy can be written in the form

$$[E - (E_{o,A} + E_{o,B})] = E_{o,AB} + \sum_{k=1}^{\infty} E_k ,$$

there appears a zeroth order energy correction, which corresponds to a purely coulombic interaction between the charge desnsities $\rho_A = \Psi_A \Psi_A^*$ and $\rho_B = \Psi_B \Psi_B^*$.

The structure of the higher order correction terms will easily follow from the expression of E. So we have

$$E_{n,I} = \sum_{p\varepsilon I} \sum_{r=o}^{n} \sum_{s=o}^{n-r} \langle p;r;I|F_{I;n-(r+s);p} + \frac{1}{2}\omega_p^I h_I \delta_{n-(r+s),o}|I;s;p\rangle ,$$

and as one can write

$$E_{AB} = \sum_{p\varepsilon A} \omega_p^A \langle p;A|V_B + \frac{1}{2} J_B|A;p\rangle$$

$$+ \sum_{q\varepsilon B} \omega_p^B \langle q;B|V_A + \frac{1}{2} J_A|B;q\rangle$$

with $\quad V_I = - \sum_{c\varepsilon I} \frac{Z_c}{r_c}$

and $\quad J_I = \sum_{p\varepsilon I} \omega_p^I J_{I;p}$,

then

$$E_{n,AB} = \sum_{p\varepsilon A} \omega_p^A \sum_{r=o}^{n} \sum_{s=o}^{n-r} \langle p;r;A|\delta_{n-(r+s),1} V_B + \frac{1}{2} J_{B;n-(r+s)}|A;s;p\rangle$$

$$+ \sum_{q\varepsilon B} \omega_q^B \sum_{r=o}^{n} \sum_{s=o}^{n-r} \langle q;r;B|\delta_{n-(r+s),1} V_A + \frac{1}{2} J_{A;n-(r+s)}|B;s;q\rangle$$

with

$$J_{I;k} = \sum_{p\varepsilon I} \omega_p^I \sum_{r=o}^{k} J_{I;r,k-r,p} .$$

Also

$$E_{n,AB} = \sum_{r=o}^{n} \sum_{s=o}^{n-r} \{\Delta^A_{r,s,n-(r+s)} + \Delta^B_{r,s,n-(r+s)}\}$$

with

$$\Delta^A_{r,s,n-(r+s)} = \sum_{p\varepsilon A} \omega^A \langle p;r;A|V_B\, \delta_{n-(r+s),1} + \frac{1}{2} J_{B;n-(r+s)}|A;s;p\rangle$$

or

$$E_{n,AB} = \Delta_n^A + \Delta_n^B .$$

In general we can write

$$I_n^A = E_{nA} + \Delta_n^A$$

$$= \sum_{p\varepsilon A} \sum_{r=o}^{n} \sum_{s=o}^{n-r} \langle p;r;A|F_{A;n-(r+s);p}$$

$$+ [\delta_{n-(r+s),o} \frac{1}{2} \omega_p^A h_A + \delta_{n-(r+s),1} V_B + \frac{1}{2} \omega_p^A J_{B;n-(r+s)}|A;s;p\rangle.$$

In this manner

$$E_n = I_n^A + I_n^B .$$

Also

$$I_n^A = \sum_{p\varepsilon A} \varepsilon^A_{np} + \sum_p \sum_{r,s} \langle p;r;A|\delta_{n-(r+s),o} \frac{1}{2} \omega_p^A h_A + \delta_{n-(r+s),1} V_B +$$

$$+ \frac{1}{2} \omega_p^A J_{B;n-(r+s)}|\ A;s;p\rangle .$$

The first order interaction energy will then be

$$E_{1,A} = \sum_{p\varepsilon A} \{\langle p;1;A|F_{A;o;p} + \frac{1}{2} \omega_p^A h_A|A;o;p\rangle$$

$$+ <p;o;A|F_{A;o;p} + \frac{1}{2}\omega_p^A h_A|A;1;p>$$

$$+ <p;o;A|F_{A;1;p}|A;o;p>\}$$

and a similar term will be obtained for $E_{1,B}$, while for Δ_1^A one finds

$$\Delta_1^A = \sum_{p\varepsilon A} \omega_p^A\{<p;1;A|\frac{1}{2} J_{B;o}|A;o;p>$$

$$+ <p;o;A|\frac{1}{2} J_{B;o}|A;1;p>$$

$$+ <p;o;A|V_B + \frac{1}{2} J_{B;1}|A;o;p>\}$$

and similarly for Δ_1^B.

In fact, one can collect terms and write

$$I_1^A = E_{1,A} + \Delta_1^A =$$

$$= \sum_{p\varepsilon A} \{<p;1;A|F_{A;o;p} + \frac{1}{2}\omega_p^A(h_A + J_{B;o})|A;o;p>$$

$$+ <p;o;A|F_{A;o;p} + \frac{1}{2}\omega_p^A(h_A + J_{B;o})|A;1;p>$$

$$+ <p;o;A|F_{A;1;p} + \omega_p^A(V_B + \frac{1}{2} J_{B;1})|A;o;p>\};$$

the first two terms are the so called induction energy and the third the dispersion term.

3.5. Electrostatic Molecular Potential

In the case where one of the molecules is changed to a positive point charge, then to the hamiltonian of A, say

$$\mathscr{H}_A = \sum_i (-\frac{1}{2}\nabla_i^2 - \sum_a \frac{Z_a}{r_{ai}} + \sum_{j<i} \frac{1}{r_{ij}})$$

should be added an attractive term of the form

$$U_H = \sum_i \frac{-1}{r_{iH}}$$

where r_{iH} is the distance of electron i to the charge. So U_H acts itself as a perturbative monoelectronic potential directly. The J_I terms in the energy expression will disappear and we can write

$$E = \sum_p \{<p|\omega_p h + \sum_q (\alpha_{pq} J_q - \beta_{pq} K_q)|p>$$

$$+ <p|\omega_p U_H|p>\} .$$

There still remains in the correction terms a zeroth order correction to the interaction energy,

$$E_{o,AH} = \sum_p \omega_p <p,o|U_H|o,p> ,$$

which is no more than the electrostatic molecular potential of A. The nth order terms can be written as

$$E_{n,AH} = \sum_p \sum_{r=o}^{n} \sum_{s=o}^{n-r} <p;r|Y_{n-(r+s);p}|s;p>$$

where

$$Y_{\ell;p} = \delta_{\ell o}\omega_p h + \delta_{\ell 1}\omega_p U_H + \sum_{r=o}^{\ell} X_{r,\ell-r;i} .$$

So, the first order correction will be

$$E_{1;AH} = \sum_p \{<p;1|Y_{o;p}|o;p> + <p;o|Y_{1;p}|o;p> + <p;o|Y_{o;p}|1;p>\}$$

with

$$Y_{o;p} = \omega_p h + X_{oo;i} = F_{o;i} + \frac{1}{2}\omega_p h$$

and

$$Y_{1;p} = \omega_p U_H + X_{1o;i} + X_{o1;i} .$$

This correction will account, as in the general case, for the induction and dispersion energies.

VII. General Theory for Two and Three Electron Systems

We will study here the Fock operators formalism for two and three electron systems, in order to deal with comparatively easy cases. Fock operators for generalized situations can be found in Appendix C.

1. Two Electron Systems

1.1. Singlet States

In this section we shall take into account all possible configurations where the available electrons fill one or two different MO's. In the first case the wavefunction will be written as

$$D(i) = (2)^{-1/2} \, \mathrm{Det} \begin{vmatrix} \phi_i(1) & \bar{\phi}_i(1) \\ \phi_i(2) & \bar{\phi}_i(2) \end{vmatrix}$$

$$= (2)^{-1/2} \, [\phi_i(1)\phi_i(2)][\alpha(1)\beta(2) - \beta(1)\,\alpha(2)]$$

and in the second case:

$$S(i,j) = \frac{1}{2} \{ \mathrm{Det} \begin{vmatrix} \phi_i(1) & \bar{\phi}_j(1) \\ \phi_i(a) & \bar{\phi}_j(2) \end{vmatrix} - \mathrm{Det} \begin{vmatrix} \bar{\phi}_i(1) & \phi_j(1) \\ \bar{\phi}_i(2) & \phi_j(2) \end{vmatrix} \}$$

$$= \frac{1}{2} \, [\phi_i(1)\phi_j(2) + \phi_i(2)\phi_j(1)][\alpha(1)\beta(2) - \beta(1)\alpha(2)].$$

From these two expressions a unique wavefunction can be obtained

$$\Phi(i,j) = \alpha_{ij}[\phi_i(1)\phi_j(2) + \phi_i(2)\phi_j(1)] \; \Theta(1,2)$$

with

$$\alpha_{ii} = \frac{1}{2}(2)^{-1/2} \quad \text{and} \quad \alpha_{ij} = \frac{1}{2}$$

also

$$\Theta(1,2) = [\alpha(1)\beta(2) - \alpha(2)\beta(1)].$$

If $\mathcal{H}$ is the two electron hamiltonian and we search for the matrix elements of $\mathscr{H}$ with respect to the basis set $\Phi(i,j)$, one should take into account the integral

$$\int \Theta^*(1,2)\ \Theta(1,2)\ d\sigma(1)\ d\sigma(2) = 2,$$

so if

$$\Gamma(i,j) = \alpha_{ij}[\phi_i(1)\phi_j(2) + \phi_i(2)\phi_j(1)] \ ,$$

then

$$<\Phi(i,j)|\ \mathscr{H}\ |\Phi(k,\ell)> = 2\ <\Gamma(i,j)|\ \mathscr{H}|\Gamma(k,\ell)> \ .$$

Using atomic units

$$\mathscr{H} = h(1) + h(2) + \frac{1}{r_{12}}$$

with

$$h(i) = -\frac{1}{2}\Delta_i - \sum_A \frac{Z_A}{r_{iA}} \ .$$

In this case if $\beta_{ii} = (2)^{-1/2}$ and $\beta_{ij} = 1$, then

$$<\Phi(i,j)|\ \mathscr{H}|\Phi(k,\ell)> =$$

$$\beta_{ij}\beta_{k\ell}\{\delta_{j\ell}(i|k) + \delta_{i\ell}(j|k) + \delta_{jk}(i|\ell) + \delta_{ik}(j|\ell)$$

$$+ (ik|j\ell) + (i\ell|jk)\}$$

with $(ik|j\ell) = \int \phi_i^*(1)\ \phi_k(1)\ \frac{1}{r_{12}}\ \phi_j^*(2)\ \phi_\ell(2)\ dV_1\ dV_2$

and $(i|k) = \int \phi_i^*(1)\ h(1)\ \phi_k(1)\ dV_1$.

The total variational wavefunction can be constructed as

$$\Psi(1,2) = \sum_i \sum_{j\geq i} c_{ij}\Phi(i,j)$$

or

$$\Psi(1,2) = \sum_i \sum_j d_{ij}\Phi(i,j),$$

using the fact that $\phi(i,j) = \phi(j,i)$, and defining

$$d_{ij} = (2)^{-1/2} c_{ij} \text{ and } d_{ii} = c_{ii}$$

with the symmetric property $d_{ij} = d_{ji}$.
Then the total energy will be given by

$$E = \sum_{i,j} \sum_{k,\ell} g_{ij} g_{k\ell} \{ \delta_{j\ell}(i|k) + \delta_{i\ell}(j|k) + \delta_{jk}(i|\ell) + \delta_{ik}(j|\ell)$$

$$+ (ik|j\ell) + (i\ell|jk) \} ,$$

with $g_{ij} = d_{ij}\beta_{ij} = (2)^{-1/2} c_{ij}$, and $g_{ij} = g_{ji}$.

Each term in the sum can be further simplified:

a) $$\sum_{i,j} \sum_{k,\ell} g_{ij} g_{k\ell} \delta_{i\ell}(j|k) = \sum_i \sum_j \sum_k g_{ij} g_{ki}(j|k)$$

$$= \sum_i \sum_j \sum_k g_{ji} g_{kj}(i|k)$$

b) $$\sum_{i,j} \sum_{k,\ell} g_{ij} g_{k\ell} \delta_{jk}(i|\ell) = \sum_i \sum_j \sum_k g_{ij} g_{j\ell}(i|\ell)$$

$$= \sum_i \sum_j \sum_k g_{ij} g_{jk}(i|k)$$

c) $$\sum_{i,j} \sum_{k,\ell} g_{ij} g_{k\ell} \delta_{ik}(j|\ell) = \sum_i \sum_j \sum_\ell g_{ij} g_{i\ell}(j|\ell)$$

$$= \sum_i \sum_j \sum_k g_{ji} g_{jk}(i|k)$$

d) $$\sum_{i,j} \sum_{k,\ell} g_{ij} g_{k\ell} \delta_{j\ell}(i|k) = \sum_i \sum_j \sum_k g_{ij} g_{kj}(i|k).$$

Since the coefficients $\{g_{ij}\}$ are the elements of a symmetric matrix, the four monoelectronic terms will be the same. Both bielectronic terms are also equal,

$$\sum_{i,j} \sum_{k,\ell} g_{ij} g_{k\ell}(i\ell|jk) = \sum_{i,j} \sum_{k,\ell} g_{ij} g_{k\ell}(ik|j\ell),$$

and the final energy expression is

$$E = 2 \sum_i \sum_j \sum_k \sum_\ell g_{ij} g_{k\ell} [2\ \delta_{j\ell} (i|k) + (ik|j\ell)].$$

The energy variation gives

$$\delta E = 2 \sum_{i,j} \sum_{k,\ell} g_{ij} g_{k\ell} [2\ \delta_{j\ell} (\delta i|k) + \langle \delta i | V_{j\ell} | k \rangle + \langle \delta j | V_{ik} | \ell \rangle]$$

+ complex conjugate terms.

Since

$$\sum_{i,j} \sum_{k,\ell} g_{ij} g_{k\ell} \langle \delta j | V_{ik} | \ell \rangle = \sum_{i,j} \sum_{k,\ell} g_{ij} g_{\ell k} \langle \delta i | V_{j\ell} | k \rangle ,$$

the final expression for the varied energy can be written as

$$\delta E = 4 \sum_{i,j} \sum_{k,\ell} g_{ij} g_{k\ell} [\delta_{j\ell} (\delta i|k) + (\delta i|V_{j\ell}|k)] + \text{c.c. terms.}$$

So the Fock Operators will be

$$F_{ik} = \sum_j \sum_\ell g_{ij} g_{k\ell} [\delta_{j\ell}\ h + V_{j\ell}] ,$$

and the energy expression in terms of the defined Fock operators will be

$$E = 2 \sum_i \sum_k \langle i | F_{ik} + \omega_{ik}\ h | k \rangle$$

with

$$\omega_{ik} = \sum_j g_{ij} g_{kj} .$$

If only paired excitations are considered, that is, if only the functions D(i,j) are taken into account, it is found that

$$E_D = 2 \sum_i \sum_k g_{ii} g_{kk} [2\ \delta_{ik} (i|k) + (ik|ik)]$$

with diagonal Fock operators like

$$F_{ij} = g_{ii}^2\, h + g_{ii} \sum_k g_{kk} K_k \ ,$$

as already found in section 3.V.
If only the single excitation wavefunctions occur in the MC wavefunction, then

$$E_S = 2 \sum_{i,k} \sum_{j,\ell} g_{ij} g_{k\ell} [2\ \delta_{j\ell} (i|k) + (ik|j\ell)]\ (1 - \delta_{ij})(1 - \delta_{k\ell}),$$

and in this case the Fock operators will be

$$F_{ik} = \sum_j \sum_\ell \rho_{ij} \rho_{k\ell} \{\delta_{j\ell}\ h + V_{j\ell}\}$$

where

$$\rho_{ij} = g_{ij} (1 - \delta_{ij}).$$

1.2. Triplet States

We use here all possible distributions of two electrons with the same spin in any pair of MO's. The determinantal wavefunction for each case can be written as

$$T(i,j) = (2)^{-1/2} \ \mathrm{Det} \begin{vmatrix} \phi_i(1) & \phi_j(1) \\ \phi_i(2) & \phi_j(2) \end{vmatrix}$$

$$= (2)^{-1/2} \{\phi_i(1)\ \phi_j(2) - \phi_i(2)\ \phi_j(1)\}\{\alpha(1)\ \alpha(2)\} \ .$$

The Hamiltonian matrix elements will be in this case

$$\langle T(i,j) | \mathcal{H} | T(k,\ell) \rangle =$$

$$\tfrac{1}{2} \{ \delta_{j\ell} (i|k) + \delta_{ik} (j|\ell) - \delta_{jk} (i|\ell) - \delta_{i\ell} (j|k)$$

$$+ (ik|j\ell) - (i\ell|jk) \} \ .$$

Because $T(i,j) = - T(j,i)$, and $T(i,i)$ does not exist then the total wavefunction can be written as

$$\Psi = \sum_i \sum_{j>i} c_{ij} T(i,j) = \sum_i \sum_j d_{ij} T(i,j)$$

with $d_{ii} = 0$; $d_{ij} = (2)^{-1/2} c_{ij}$ and $d_{ij} = -d_{ji}$, or $\{d_{ij}\}$ forms a skew symmetric matrix.

The total energy expression, after simplification of terms, following the same procedure as the singlet case, is

$$E = 2 \sum_{i,j} \sum_{k,\ell} d_{ij} d_{k\ell} \{2\ \delta_{j\ell} (i|k) + (ik|j\ell)\}.$$

A formula identical to that in the singlet case arises, except the coefficients of the total wavefunction are different. Of course, equal Fock operators as in the singlet case, with the appropriate coefficients, shall be used here.

2. Three Electron Systems

2.1. Doublet States. Case A

In the first case we take into account all the possible configurations which arise from distribution of the three electrons in the MO's with a paired double and a single occupied MO's.

In any of these situations the wavefunctions will have the form:

$$\Phi(i,k) = (6)^{-1/2} \text{ Det} \begin{vmatrix} i(1) & \bar{i}(1) & k(1) \\ i(2) & \bar{i}(2) & k(2) \\ i(3) & \bar{i}(3) & k(3) \end{vmatrix}.$$

The hamiltonian matrix element will be

$$\langle\Phi(i,k)|\mathcal{H}|\Phi(j,\ell)\rangle =$$

$$(2\ \delta_{ij}\delta_{k\ell} - \delta_{kj}\delta_{i\ell})(i|j) + \delta_{ij}(k|\ell)$$

$$+ \delta_{ij}[2(ij|k\ell) - (kj|i\ell)] - \delta_{i\ell}(ij|kj)$$

$$- \delta_{kj}(i\ell|ij) + \delta_{k\ell}(ij|ij).$$

The MC wavefunction can be written as

$$\Psi = \sum_i \sum_k (1 - \delta_{ik})\ d_{ik}\ \Phi(i,k) = \sum_i \sum_k c_{ik}\ \Phi(i,k),$$

and the electronic energy:

$$E = \sum_{i,k} [\omega_{ik}(i|k) + \sum_{\ell}\{\alpha_{ik}^{\ell}(ik|\ell\ell) - \beta_{ik}^{\ell}(i\ell|k\ell)\}]$$

with

$$\omega_{ik} = \omega_{ki} = \sum_{j} 2(\delta_{ik}c_{ij}^2 + \delta_{kj}\, c_{ik}\, c_{ki} + c_{ji}\, c_{jk})$$

$$\alpha_{ik}^{\ell} = \alpha_{ki}^{\ell} = 2\, c_{\ell k}\, c_{\ell i}$$

$$\beta_{ik}^{\ell} = \beta_{ki}^{\ell} = c_{\ell k}\, c_{\ell i} + c_{ik}\, c_{\ell i} + c_{ki}\, c_{\ell k} - \delta_{ik} \sum_{j} c_{ij}\, c_{\ell j},$$

and the Fock operators will be

$$F_{ik} = \omega_{ik}\, h + \sum_{j,\ell} (\alpha_{ik}^{j\ell}\, V_{j\ell} - \beta_{ik}^{j\ell}\, X_{j\ell}),$$

with the new parameters

$$\alpha_{ik}^{j\ell} = \delta_{j\ell}\, \alpha_{ik}^{\ell} + \delta_{ik}\, \alpha_{j\ell}^{i}$$

and

$$\beta_{ik}^{j\ell} = \delta_{j\ell}\, \beta_{ik}^{\ell} + \delta_{ik}\, \beta_{j\ell}^{i} .$$

It is easy to show that $F_{ik}^{+} = F_{ki}$ and that $E = \frac{1}{2} \sum_{i} \sum_{k} \langle i|F_{ik} + \omega_{ik}\, h|k\rangle$.

2.2. Doublet States. Case B

In this case all the possible ways to arrange three electrons in all the available MO's will be considered. That is, we can have the situation where Case A is mixed together with three singly occupied MO's. The monoconfigurational wavefunction can be written

$$\Phi(i,j,k) = \frac{1}{2}(3)^{-1/2}\{[i\ \bar{j}\ k] - [\bar{i}\ j\ k]\} ,$$

with the symbol $[i\ \bar{j}\ k]$ representing a determinant like the one in case A. After a tedious but straightforward computation, following the same pattern as in the cases already studied, and taking into account that the total wavefunction can be written as

$$\Psi = \sum_i \sum_j \sum_k (1 - \delta_{ik})(1 - \delta_{jk})\; c_{ijk}\; \Phi(i,j,k)$$

$$= \sum_i \sum_j \sum_k d^k_{ij}\; \Phi(i,j,k),$$

using $\Phi(i,j,k) = \Phi(j,i,k)$, then the coefficients will have the symmetry conditions

$$d^k_{ij} = d^k_{ji},$$

and the energy expression will be

$$E = \sum_{i,j,k} \sum_{p,q,r} d^k_{ij}\; d^r_{pq}\; \langle \Phi(i,j,k) | \mathscr{H} | \Phi(p,q,r) \rangle$$

$$= \sum_{i,p} (\omega_{ip}(i|p) + \sum_{k,r} \{\alpha^{kr}_{ip}(ip|kr) - \beta^{kr}_{ip}(ir|kp)\})$$

with

$$\omega_{ip} = \omega_{pi} = 2 \sum_{j,k} \{(2\, d^k_{ij}\, d^k_{pj} - d^k_{ij}\, d^k_{pq}) +$$

$$d^i_{kj}\, d^p_{kj} - (d^k_{ij}\, d^p_{kj} + d^k_{pj}\, d^i_{kj})\}$$

$$\alpha^{kr}_{ip} = \alpha^{rk}_{pi} = 2 \sum_j (2\, d^k_{ij}\, d^r_{pj} + d^j_{ik}\, d^j_{pr})$$

and

$$\beta^{kr}_{ip} = \beta^{rk}_{pi} = 2 \sum_j (d^k_{ij}\, d^r_{pj} + d^k_{ij}\, d^j_{pr} + d^r_{pj}\, d^j_{ik}).$$

In this case the Fock operators will be

$$F_{ip} = \omega_{ip}\, h + \sum_{k,r} (\lambda^{kr}_{ip}\, V_{kr} - \mu^{kr}_{ip}\, X_{kr})$$

with

$$\lambda_{ip}^{kr} = \lambda_{pi}^{rk} = \alpha_{ip}^{kr} + \alpha_{kr}^{ip}$$

and

$$\mu_{ip}^{kr} = \mu_{pi}^{rk} = \beta_{ip}^{kr} + \beta_{kr}^{ip} .$$

2.3. Quadruplet States

If all the ways to distribute three electrons in the MO's forming a quadruplet state are considered, the generic wavefunction can be written as

$$\Phi(i,j,k) = (6)^{-1/2} [i,j,k] ,$$

and the total wavefunction becomes

$$\Psi = \sum_{i,j,k} (1 - \delta_{ij})(1 - \delta_{ik})(1 - \delta_{jk}) \; c_{ijk} \; \Phi(i,j,k)$$

$$= \sum_{i,j,k} d_{ijk} \; \Phi(i,j,k) .$$

Using the fact that

$$\Phi(i,j,k) = \Phi(j,k,i) = \Phi(k,i,j)$$

$$= - \Phi(j,i,k) = - \Phi(k,j,i) = - \Phi(i,k,j)$$

it follows that

$$d_{ijk} = d_{jki} = d_{kij} = - d_{jik} = - d_{kji} = - d_{ikj} ,$$

so the total energy adopts the form

$$E = \sum_{i,p} (\omega_{ip}(i|p) + \sum_{k,r} \alpha_{ip}^{kr} \{(ip|kr) - (ir|kp)\})$$

with

$$\omega_{ip} = \omega_{pi} = 18 \sum_{j,k} d_{ijk} \; d_{pjk}$$

and

$$\alpha_{ip}^{kr} = 9 \sum_j d_{jik}\, d_{jpr} ,$$

with the properties

$$\alpha_{ip}^{kr} = \alpha_{pi}^{rk} = \alpha_{kr}^{ip} = \alpha_{rk}^{pi}$$

$$= - \alpha_{ir}^{kp} = - \alpha_{pk}^{ri} = - \alpha_{kp}^{ir} = - \alpha_{ri}^{pk} .$$

An alternative energy expression is

$$E = \sum_{i,p} \{\omega_{ip}(i|p) + 2 \sum_{k,r} \alpha_{ip}^{kr} (ip|kr)\} ,$$

then the Fock operators will be in this case

$$F_{ip} = \omega_{ip}\, h + 4 \sum_{k,r} \alpha_{ip}^{kr} V_{kr} .$$

VIII. Approximate SCF Theories

It is well known that the main SCF problem, if one leaves out the SCF solution itself, lies in the calculation and handling of bielectronic integrals needed when the LCAO forms of Fock operators are constructed. The problem can be considered alleviated and solved within the present knowledge through the efficient algorithms in use. However, if one looks ahead in time, then in some cases, such as the study of huge biological molecules, the use of approximate methods can still be useful. When dealing with approximate SCF procedures one needs to divide the methodology into two separate units: non-empirical and empirical methods.

By non-empirical methods we mean procedures which use only integral approximations, while the term empirical is reserved for methods that, besides these simplifications, also use experimental parameters and empirical constants to allow further computational simplifications. In this section we will briefly analyze these two possibilities.

1. Atomic Orbital Representation

1.1. Introduction

Since they were published, a wide number of papers have dealt with the problem of finding better and more meaningful approaches to Mulliken's approximation and population analysis. A careful study of the existing literature shows how the problem has been treated from different points of view. Moreover the approximation of molecular integrals has often been studied by taking orthogonal or orthogonalized AO basis sets as the initial step.

In this section we will show how both problems are closely related and due to the mathematical properties of AO generated functional spaces. This close relationship will serve to obtain direct integral approximations without requiring compulsive orthogonality on basis sets. Partitioning of the expected values of operators among atomic contributions comes as a natural consequence of this approach.

1.2. Representation of AO's

Let's suppose that an atomic basis set associated with some molecular system is centered on various points of space named A, B ... ; a partition of the whole basis set can be made taking the AO subsets

centered on each point, namely $\{\chi_{\mu_A}\}$ $\{\chi_{\nu_B}\}$, ...
Using the properties of the AO functional space we can express each orbital on center A as a linear combination of orbitals centered on B, so one can write

$$\chi_{\mu_A} \quad \sum_{\lambda B} a_{\lambda_B \mu_A} \chi_{\lambda_B} , \qquad (1.\mathrm{VIII})$$

where $\{a_{\lambda_B \mu_A}\}$ is a set of appropriate coefficients.
As the AO subsets on each center are linearly independent, the representation (1.VIII) is unique and the coefficients will be given by the expression

$$a_{\lambda_B \mu_A} = \sum_{\beta_B} S^{(-1)}_{\lambda_B \beta_B} S_{\beta_B \mu_A} \qquad (2.\mathrm{VIII})$$

where $S^{(-1)}_{\lambda_A \beta_B}$ is an element of the inverse metric matrix of the AO subset centered on B and

$$S_{\beta_B \mu_A} = \int \chi^*_{\beta_B} \chi_{\mu_A} \, dV$$

is an element of the metric matrix of the whole molecular AO basis set. We will have, in general, combining equations (1.VIII) and (2.VIII):

$$\chi_{\mu_A} = \sum_{\lambda_B} \sum_{\beta_B} S^{(-1)}_{\lambda_B \beta_B} S_{\beta_B \mu_A} \chi_{\lambda_B} . \qquad (3.\mathrm{VIII})$$

Ruedenberg's approximation is a very particular case of this formalism, and results from (3.VIII) by taking the subset $\{\chi_{\lambda_B}\}$ as orthonormalized, so that

$$S_{\lambda_B \beta_B} = S^{(-1)}_{\lambda_B \beta_B} = \delta_{\lambda_B \beta_B} , \qquad (4.\mathrm{VIII})$$

and (3.VIII) transforms into

$$\chi_{\mu_A} = \sum_{\lambda_B} S_{\lambda_B \mu_A} \chi_{\lambda_B} \qquad (5.\mathrm{VIII})$$

which is an obvious and very well known result. It should be stressed that (5.VIII) cannot be used unless conditions (4.VIII) are fulfilled.

1.3. Representation of Charge Density: Mulliken's Gross Atomic Populations as a Natural Way of Charge Partitioning

Once equation (1.VIII) is adopted as a working tool, a charge distribution involving two AO's on different centers,

$$\rho_{\mu_A \nu_B} \doteq \chi_{\mu_A} \chi^{*}_{\nu_B} , \tag{6.VIII}$$

can be easily obtained by averaging the representation of χ_{μ_A} in terms of the set $\{\chi_{\lambda_B}\}$ and the corresponding representation of χ_{ν_B} in terms of $\{\chi_{\lambda_A}\}$. So one can write

$$\rho_{\mu_A \nu_B} = \frac{1}{2} \sum_{\lambda_B} \sum_{\beta_B} S^{(-1)}_{\lambda_B \beta_B} S_{\beta_B \mu_A} \chi_{\lambda_B} \chi^{*}_{\nu_B} \tag{7.VIII}$$

$$+ \frac{1}{2} \sum_{\lambda_A} \sum_{\alpha_A} S^{(-1)}_{\lambda_A \alpha_A} S_{\alpha_A \nu_B} \chi_{\mu_A} \chi_{\lambda_A} . \tag{8.VIII}$$

The density function of the molecule can be written as

$$\rho = \sum_{A} \sum_{B} \sum_{\mu_A} \sum_{\nu_B} D_{\mu_A \nu_B} \rho_{\mu_A \nu_B} ,$$

where $D_{\mu_A \nu_B}$ is an element of the first order density matrix. Using (7.VIII) and (8.VIII) one finally obtains:

$$\rho = \sum_{A} \sum_{\mu_A} \sum_{\lambda_A} T_{\mu_A \lambda_A} \chi_{\mu_A} \chi^{*}_{\lambda_A} , \tag{9.VIII}$$

with

$$T_{\mu_A \lambda_A} = \frac{1}{2} \sum_{\sigma_A} \{ (S^{(-1)}_{\mu_A \sigma_A} [\sum_{B} \sum_{\nu_B} S_{\sigma_A \nu_B} D_{\nu_B \lambda_A}] \tag{10.VIII}$$

$$+ [\sum_{B} \sum_{\nu_B} D_{\mu_A \nu_B} S_{\nu_B \sigma_A}] S^{(-1)}_{\sigma_A \lambda_A}) \} .$$

As a consequence of these expressions, if we wish to find an adequate representation of the total number of electrons in the molecule, it will be sufficient to compute the integral

$$N = \int \rho \, dV = \sum_{A} \sum_{\mu_A} \sum_{\lambda_A} T_{\mu_A\lambda_A} S_{\mu_A\lambda_A} . \tag{11.VIII}$$

It follows that a natural partition of the electronic charge will be

$$N = \sum_{A} Q_A , \tag{12.VIII}$$

with

$$Q_A = \sum_{\mu_A} \sum_{\lambda_A} T_{\mu_A\lambda_A} S_{\mu_A\lambda_A} . \tag{13.VIII}$$

But definition (10.VIII) along with some straightforward manipulation of (13.VIII) gives

$$Q_A = \sum_{\mu_A} \sum_{B} \sum_{\nu_B} D_{\mu_A\nu_B} S_{\nu_B\mu_A} , \tag{14.VIII}$$

which is the Mulliken's definition of gross atomic population. Also Mulliken's orbital populations Q_{μ_A} are simply

$$Q_{\mu_A} = \sum_{\lambda_A} T_{\mu_A\lambda_A} S_{\mu_A\lambda_A} = \sum_{B} \sum_{\nu_B} D_{\mu_A\nu_B} S_{\nu_B\mu_A} . \tag{15.VIII}$$

These results come from the fact that representation (1.VIII) leaves the whole metric matrix invariant.

In this manner we arrive at the following conclusions:

1) The representation (1.VIII) associated to the charge distribution (7.VIII) leaves invariant the molecular charge.
2) Mulliken's gross atomic populations, under representation (1.VIII), are a natural way to distribute molecular charge between molecular centers or atomic orbitals.
3) Mulliken's populations will strongly depend on the nature of the AO subsets through the matrix elements $T_{\mu_A\lambda_A}$.

   This is a well known result, but nothing more can be said if one takes (1.VIII) as a working hypothesis.
4) Any other choice of charge partition will become arbitrary if (1.VIII)

is assumed.

1.4. Expectation Values of Monoelectronic Operators

The expectation value of a monoelectronic hermitean operator Ω can be written as

$$\langle\Omega\rangle = \sum_A \sum_B \sum_{\mu_A} \sum_{\nu_B} D_{\mu_A\nu_B} \langle\mu_A|\Omega|\nu_B\rangle \ , \qquad (16.\text{VIII})$$

where $D_{\mu_A\nu_B}$ is the first order density matrix, and

$$\langle\mu_A|\Omega|\nu_B\rangle = \int \rho_{\mu_A\nu_B} \ \Omega \ dV \ . \qquad (17.\text{VIII})$$

Then, substituting the representation of the charge distribution $\rho_{\mu_A\nu_B}$ given by equation (7.VIII) into equation (17.VIII), and using equation (15.VIII) gives:

$$\langle\Omega\rangle = \sum_A \sum_{\mu_A} \sum_{\lambda_A} T_{\mu_A\lambda_A} \ (\mu_A|\Omega|\lambda_A) \ , \qquad (18.\text{VIII})$$

where only one center charge distribution is involved. Furthermore, a natural partitioning of $\langle\Omega\rangle$ into atomic terms is given by

$$\langle\Omega\rangle = \sum_A \langle\Omega\rangle_A \ , \qquad (19.\text{VIII})$$

with

$$\langle\Omega\rangle_A = \sum_{\mu_A} \sum_{\lambda_A} T_{\mu_A\lambda_A} \ (\mu_A|\Omega|\lambda_A) \ . \qquad (20.\text{VIII})$$

Equation (18.VIII) should be taken as an approximation to $\langle\Omega\rangle$, which will converge to the exact multicentric expression (16.VIII) as the AO basis increases in size.

The expression for $\langle\Omega\rangle$ obtained from equation (20.VIII) by applying Mulliken's approximation uses an average of the monocentric integrals as follows:

$$(\mu_A|\Omega|\lambda_A) \simeq \frac{1}{2} S_{\mu_A\lambda_A} \{(\mu_A|\Omega|\mu_A) + (\lambda_A|\Omega|\lambda_A)\} \ . \qquad (21.\text{VIII})$$

Substituting (21.VIII) into (20.VIII) and rearranging terms one finds

$$<\Omega> \simeq \sum_{A} \sum_{\mu_A} Q_{\mu_A} <\mu_A|\Omega|\mu_A> . \tag{22.VIII}$$

In contrast to (20.VIII), which can be used without further precautions in any case, Mulliken's expression (22.VIII) only will give adequate answers if used in an orthogonal (CNDO, INDO, etc.) atomic subset framework. This occurs because (22.VIII) can also be obtained approximately from (20.VIII) by supposing conditions (4.VIII) hold. In these circumstances the matrix elements (10.VIII) will become

$$T^{\circ}_{\mu_A\lambda_A} = \frac{1}{2} \sum_{B} \sum_{\nu_B} \{S_{\mu_A\nu_B} D_{\nu_B\lambda_A} + D_{\mu_A\nu_B} S_{\nu_B\lambda_A}\} , \tag{23.VIII}$$

which are the elements of the Chirgwin-Coulson matrix. But since by construction, the set $\{\chi_{\mu_A}\}$ is orthogonal, one can write

$$(\mu_A|\Omega|\lambda_A) \simeq \delta_{\mu_A\lambda_A} (\mu_A|\Omega|\mu_A) \tag{24.VIII}$$

so that (18.VIII) will take the form

$$<\Omega> \sim \sum_{A} \sum_{\mu_A} T^{\circ}_{\mu_A\mu_A} (\mu_A|\Omega|\mu_A) . \tag{25.VIII}$$

However, Mulliken's orbital populations are precisely the diagonal elements of Chirgwin-Coulson matrix, and as a result (25.VIII) will coincide with (22.VIII). Depending on the nature of the operator Ω, equations (24.VIII) will be exact.

1.5. SCF Theory and Energy Partitioning into Atomic Contributions

In a general SCF framework the electronic energy can be written as (see section 1.IX):

$$E = \sum_{i} \sum_{j} \{\omega_{ij} h_{ij} + \sum_{k} \sum_{\ell} \alpha_{ij}^{k\ell} (ij|k\ell)\} ,$$

where the summations run over the active MO; ω_{ij} and $\alpha_{ij}^{k\ell}$ are parameters attached to the wavefunction of the considered state; and h_{ij}, $(ij|k\ell)$ are the core and bielectronic integrals, respectively. Variation of E gives the Fock operators

$$F_{ij} = \frac{1}{2}\,\omega_{ij} h + \sum_k \sum_\ell \alpha_{ij}^{k\ell}\, V_{k\ell}$$

where h is the core operator and $V_{k\ell}$ a Coulomb operator such as

$$(ij|k\ell) = (i|V_{k\ell}|j)\ .$$

The LCAO expression of F_{ij} can be written as

$$F_{ij,\mu_A\nu_B} = \frac{1}{2}\,\omega_{ij} h_{\mu_A\nu_B} + \sum_C \sum_D \sum_{\lambda_C} \sum_{\sigma_D} A_{ij,\lambda_C\sigma_D} (\mu_A\nu_B|\lambda_C\sigma_D)\ ,$$

where $A_{ij,\lambda_C\sigma_D}$ is an element of a generalized density matrix defined as

$$A_{ij,\lambda_C\sigma_D} = \sum_k \sum_\ell \alpha_{ij}^{k\ell}\, C_{k\lambda_C}\, C_{\ell\sigma_D}\ ,$$

and $\{C_{k,\lambda}\}$ are the LCAO coefficients of the kth MO. Then, using the expression of the density distributions (7.VIII) one finds

$$F_{ij,\mu_A\nu_B} = \frac{1}{4}\,\omega_{ij}\, \{\sum_{\alpha_A} a_{\alpha_A\nu_B}\, h_{\mu_A\alpha_A} + \sum_{\beta_B} a_{\beta_B\mu_A}\, h_{\beta_B\nu_B}\}$$

$$+ \frac{1}{2} \sum_C \sum_{\lambda_C} \sum_{\gamma_C} T_{ij,\lambda_C\gamma_C} \{\sum_{\alpha_A} a_{\alpha_A\nu_B} (\mu_A\alpha_A|\lambda_C\gamma_C) + \sum_{\beta_B} a_{\beta_B\mu_A} (\beta_B\nu_B|\lambda_C\gamma_C)\}$$

where $T_{ij,\lambda_C\gamma_C}$ has the same meaning as definition (10.VIII) but with the first order density matrix elements substituted by the elements of matrices A_{ij}.

By defining for each center the matrix elements

$$G_{ij,\mu_A\alpha_A} = \frac{1}{2}\,\omega_{ij} h_{\mu_A\alpha_A} + \frac{1}{2} \sum_C \sum_{\lambda_C} \sum_{\gamma_C} T_{ij,\lambda_C\gamma_C} (\mu_A\alpha_A|\lambda_C\gamma_C)$$

the Fock matrices can be written as

$$F_{ij,\mu_A\nu_B} = \frac{1}{2}\, \{\sum_{\alpha_A} a_{\alpha_A\nu_B}\, G_{ij,\mu_A\alpha_A} + \sum_{\beta_B} a_{\beta_B\mu_A}\, G_{ij,\beta_B\nu_B}\}\ ,$$

and the whole SCF process at any level will depend only on bicentric

terms.
Using the definition

$$R_{ij,\mu_A\alpha_A} = \frac{1}{2}\,\omega_{ij} h_{\mu_A\alpha_A} + F_{ij,\mu_A\alpha_A}$$

for the matrix elements, the energy can be rewritten as

$$E = \frac{1}{2}\sum_i \sum_A \sum_{\mu_A} \sum_j \sum_B \sum_{\nu_B} C_{i\mu_A} C_{j\nu_B} \{\sum_{\alpha_A} a_{\alpha_A\nu_B} R_{ij,\mu_A\alpha_A} + \sum_{\beta_B} a_{\beta_B\mu_A} R_{ij,\beta_B\nu_B}\}$$

$$= \sum_A \sum_{\mu_A} \sum_{\alpha_A} E_{\mu_A\alpha_A} = \sum_A E_A$$

which can be considered as a natural energy partition between atomic contributions.

1.6. General Remarks

The most relevant conclusion of the present development is the capacity of an expansion like (1.VIII) to separate the molecular observables into atomic contributions. This partition, which includes also molecular energy, is a natural one, unique due to the mathematical structure of the LCAO space representation. Any other partition becomes arbitrary in this mathematical sense. Moreover, each atomic contribution appears to have a strong dependence on the whole AO basis set. Finally, the calculated observables will tend to be more accurate with respect to the usual many center expansion as the size of the atomic basis set increases. An approximate SCF formalism can be finally set up, where only two center integrals are needed.

2. Decomposition of the Electronic Repulsion Matrix

A recently proposed idea by Linderberg and Beebe can be used to set up a systematic methodology to estimate the bielectronic integrals. The foundation of this idea lies in the fact that the bielectronic operators $\{r_{ij}^{-1}\}$ are positive definite, so that the matrix representation of this kind of operator will always be a positive definite matrix. If $\{\chi_\mu\}$ is the AO basis set for a given system, and

$$\rho_{\mu\nu}(i) = \chi_\mu(i)\chi_\nu^*(i)$$

represents the density distribution of electron $\underline{i}$ associated with AO's μ and ν, then the bielectronic integrals can be calculated from

$$(\mu\nu|\frac{1}{r}|\lambda\sigma) = \int \rho_{\mu\nu}(1)\rho_{\lambda\sigma}(2) \frac{1}{r_{12}} dv_1 dv_2 .$$

So, one can consider the set $\{\rho_{\mu\nu}(i)\}$ as a basis generating a $(n^2 + n)/2$ dimensional space, n being the dimension of the original AO space. By rearranging the index pairs (μ,ν) in a unique index set

$$I = \mu + \nu(\nu - 1)/2 \qquad (\mu \leq \nu) ,$$

one can write an equivalent representation of the bielectronic integrals as

$$V_{IJ} = (I|\frac{1}{r}|J) = (\mu\nu|\frac{1}{r}|\lambda\sigma) .$$

Due to the structure of this definition, the matrix V will be hermitean and positive definite due to the nature of the represented operator. Since V is positive definite, there will always exist an upper triangular matrix T, which fulfills the Cholesky's decomposition of V:

$$V = T^{+}T .$$

The elements of T can be easily evaluated in terms of the elements of V, and a simple and stable algorithm for doing this is known. In fact one can write

$$T_{II} = (V_{II} - (1 - \delta_{I1}) \sum_{K=1}^{I-1} |T_{KI}^2|^2)^{1/2}$$

and

$$T_{IJ} = T_{II}^{-1} (V_{IJ} - (1 - \delta_{I1}) \sum_{K=1}^{I-1} T_{KJ}^{*} T_{KI})$$

with $J > I$. Therefore, the computation should be carried out row by row, starting at the first. Also

$$V_{IJ} = \sum_{K=1}^{I} T_{KI}^{*} T_{KJ}, \quad (I \leqslant J) .$$

The practical use of the Cholesky's decomposition of V can be attributed

to the existence of some rows in T with small elements.
Suppose the diagonal elements of V are already known, and ordered in such a way as to have

$$V_{11} \geqslant V_{22} \geqslant \dots V_{II} \geqslant \dots ,$$

then the first row of V, with elements V_{IJ}, can be computed and used to construct T_{IJ}. The procedure continues with the following rows of V and computing the rows of T, until for a given index I

$$T_{II} \leqslant \varepsilon ,$$

ε being a given sufficiently small number.
At this stage the triangular matrix T, may be given a hypermatricial structure such as

$$T = \begin{bmatrix} Y \\ \hline Z \end{bmatrix} ,$$

where Z is a rectangular matrix which contains the rows of T with $T_{II} \leqslant \varepsilon$. Then

$$V = T^{+}T = Y^{+}Y + Z^{+}Z$$

and the euclidean norm $||Z^{+}Z||_2$ can be associated with the square error incurred when one seeks to compute V approximately with

$$V \sim Y^{+}Y .$$

As one needs only some rows of V to construct Y, the method can save a considerable amount of computational time, when the norm $||Z^{+}Z||_2$ is sufficiently small.
Using this approximate form one can write

$$V_{IJ} = \sum_K Y^{*}_{KI} Y_{KJ} ,$$

with K running over a number N of terms which will always be $N \leqslant (n^2 + n)/2$. In fact this procedure is somehow related to the spectral decomposition of V:

$$V = \sum_{I}^{N} \lambda_I C_I C_I^+ + \sum_{J=N+1}^{(n^2+n)/2} \mu_J C_J C_J^+$$

where $\{C_I\}$ are the eigenvectors and $\{\lambda_I\} \cup \{\mu_J\}$ the eigenvalues of V. With the additional condition

$$\mu_J \leqslant \varepsilon ,$$

holding, one finds

$$V \sim \sum_{I}^{N} \lambda_I C_I C_I^+ \quad .$$

In hypermatricial form, if Λ is the diagonal matrix with the V eigenvalues and C the eigenvector matrix

$$V = C\Lambda C^+ ,$$

and if we consider

$$C = [C_1 \;\vdots\; C_z]$$

$$\Lambda = \begin{bmatrix} \Lambda_1 & 0 \\ 0^+ & \Lambda_z \end{bmatrix} \quad \text{with} \quad \left\{ \begin{array}{l} \Lambda_1 = \text{Diag}(\lambda_I) \\ \Lambda_z = \text{Diag}(\mu_I) \end{array} \right.$$

then

$$V = C_1\Lambda_1 C_1^+ + C_z\Lambda_z C_z^+$$

and

$$V \sim C_1\Lambda_1 C_1^+ \quad .$$

The error incurred will be related to the norm

$$||C_z\Lambda_z C_z^+||_2$$

and as $\Lambda_1 > 0$, $\Lambda_1^{1/2}$ is easily computed and one can write

$$U_1\Lambda_1^{1/2}\, C_1^+ = Y ,$$

being U_1 unitary, which, due to the full Cholesky's decomposition, can be related to the eigenvalues and vectors of V through

$$T = U\Lambda^{1/2}C^+$$

since there always exists a unitary matrix U which transforms the matrix $\Lambda^{1/2}C^+$ into a triangular matrix.

3. Empirical Approximate Methods

In the current empirical approximate methodology, the main simplification consists of taking into account only the valence orbitals in the SCF process, and considering the molecular core included in some empirical manner in the monoelectronic integrals. This restructuring of the molecular hamiltonian can also be done without the need of other approximations about the valence part of the system, through the use of the Model Potential formalism as we will show later.

In the second place, the approximate methods adopt different degrees of complexity, as they make various simplifying assumptions about the structure of bielectronic integrals.

If one writes a tetracentric bielectronic integral involving four AO as $(\mu_A\nu_B|\lambda_C\sigma_D)$, one can suppose that:

$$\text{A)}\quad (\mu_A\nu_B|\lambda_C\sigma_D) = \delta_{AB}\,\delta_{CD}(\mu_A\nu_A|\lambda_C\sigma_C)$$

which corresponds to discarding tri and tetracentric integrals.

$$\text{B)}\quad (\mu_A\nu_A|\lambda_C\sigma_C) = \delta_{\mu\nu}\,\delta_{\lambda\sigma}(\mu_A\mu_A|\lambda_C\lambda_C)$$

which corresponds to discarding bicentric hybrid and exchange integrals.

And finally one can also make a further simplication:

$$\text{C)}\quad (\mu_A\mu_A|\lambda_C\lambda_C) = (s_As_A|s_Cs_C)\ ,$$

which corresponds to averaging the AO density functions on each center to spherical s-type orbital densities.

Simplifications like A) are called NNDO (Neglect of Diatomic Differential Overlap), those of type C) are named CNDO (Complete Neglect of Diatomic Differential Overlap). Both simplifications maintain rotational invariance of the energy, on the contrary approximation B) does not.

An intermediate form between B) and C) designed to include rotational invariance is called INDO (Intermediate Neglect of Differential Overlap), which can be described as

$$D)^{*} \quad (\mu_A \nu_A | \lambda_C \sigma_C) = \delta_{\mu\nu}\, \delta_{\lambda\sigma} (S_A S_A | S_C S_C) \text{ if } A \neq C$$

$$* \quad (\mu_A \nu_A | \lambda_A \sigma_A) \text{ are conserved.}$$

In any case, the SCF formalism will not be affected by these approximations, and only the form of the Fock operators will suffer the consequences of neglecting the integrals.
Therefore, if the general LCAO form of the Fock operators is written as

$$F_{i,\mu\nu} = \frac{1}{2}\, \omega_i h_{\mu\nu} + \sum_{\lambda} \sum_{\sigma} (A_{i,\lambda\sigma}(\mu\nu|\lambda\sigma) - B_{i,\lambda\sigma}(\mu\lambda|\nu\sigma))$$

we will have the following simplified forms:

a) <u>CNDO structure of $F_{i,\mu\nu}$</u>

Using approximations A) and B) we will have

$$F_{i,\mu\mu} = \frac{1}{2}\, \omega_i h_{\mu\mu} + \sum_{\lambda} \{A_{i,\lambda\lambda}(\mu\mu|\lambda\lambda)\} - B_{i,\mu\mu}(\mu\mu|\mu\mu)$$

and

$$F_{i,\mu\nu} = \frac{1}{2}\, \omega_i h_{\mu\nu} + B_{i,\mu\mu}(\mu\mu|\nu\nu); \text{ if } \mu \neq \nu \quad ,$$

then using C) and calling $\alpha_{AB} = (S_A S_A | S_B S_B)$

$$F_{i,\mu\mu} = \frac{1}{2}\, \omega_i h_{\mu\mu} + \sum_{B} A_{i,B}\, \alpha_{AB} - B_{i,\mu\mu}\, \alpha_{AA}; \text{ for } \mu \in A$$

with $A_{i,B} = \sum_{\lambda \in B} A_{i,\lambda\lambda}$

and for off-diagonal elements we will have

$$F_{i,\mu\mu} = \frac{1}{2}\, \omega_i h_{\mu\nu} - B_{i,\mu\mu}\, \alpha_{AB}; \text{ for } \mu \in A,\ \nu \in B\ .$$

b) <u>INDO structure of $F_{i,\mu\nu}$</u>

The off-diagonal elements of $F_{i,\mu\nu}$ with $\mu \varepsilon A$ and $\nu \varepsilon B$ have the same form, in this case, as in the CNDO approximation. The INDO approximation affects the terms with $\mu, \nu \varepsilon A$. So, using approximation D)

$$F_{i,\mu\mu} = \frac{1}{2}\omega_i h_{\mu\mu} + \sum_{\lambda\varepsilon A}\sum_{\nu\varepsilon A}(A_{i,\lambda\sigma}(\mu\mu|\lambda\sigma) - B_{i,\lambda\sigma}(\mu\lambda|\mu\sigma))$$

$$+ \sum_{B\neq A} A_{i,B}\,\alpha_{AB}\ ,$$

and the intraatomic crossed terms will be

$$F_{i,\mu\nu} = \frac{1}{2}\omega_i h_{\mu\nu} + \sum_{\lambda\varepsilon A}\sum_{\sigma\varepsilon A}(A_{i,\lambda\sigma}(\mu\nu|\lambda\sigma) - B_{i,\lambda\sigma}(\mu\lambda|\nu\sigma))$$

with $\mu \neq \nu$ and $\mu, \nu \varepsilon A$.

If a basis set of s-type and p-type orbitals is used, only the integrals of types $(\mu\mu|\lambda\lambda)$ and $(\mu\lambda|\mu\lambda)$ will survive, so we will finally have

$$F_{i,\mu\mu} = \frac{1}{2}\omega_i h_{\mu\mu} + \sum_{\lambda\varepsilon A}(A_{i,\lambda\lambda}(\mu\mu|\lambda\lambda) - B_{i,\lambda\lambda}(\mu\lambda|\mu\lambda)) + \sum_{B\neq A} A_{i,B}\,\alpha_{AB}$$

and

$$F_{i,\mu\nu} = \frac{1}{2}\omega_i h_{\mu\nu} + A_{i,\mu\nu}(\mu\nu|\mu\nu) - B_{i,\mu\nu}[(\mu\mu|\nu\nu) + (\mu\nu|\mu\nu)]\ .$$

c) <u>NDDO structure of $F_{i,\mu\nu}$</u>

Using the approximation A) we will have for $\mu, \nu \varepsilon A$

$$F_{i,\mu\nu} = \frac{1}{2}\omega_i h_{\mu\nu} + \sum_{B}\sum_{\lambda\varepsilon B}\sum_{\sigma\varepsilon B} A_{i,\lambda\sigma}(\mu\nu|\lambda\sigma)$$

$$- \sum_{\lambda\varepsilon A}\sum_{\sigma\varepsilon A} B_{i,\lambda\sigma}(\mu\lambda|\nu\sigma)$$

and for $\mu \varepsilon A$ and $\nu \varepsilon B$

$$F_{i,\mu\nu} = \frac{1}{2}\omega_i h_{\mu\nu} - \sum_{\lambda\varepsilon A}\sum_{\sigma\varepsilon B} B_{i,\lambda\sigma}(\mu\lambda|\nu\sigma)\ .$$

In all three cases studied here, the elements of the core matrix $\{h_{\mu\nu}\}$

should be evaluated according to some empirical approximations, which are irrelevant at present, since we are mainly interested in the SCF structure itself.

4. Model Potentials: Huzinaga's Approach

In the Model Potential approach, the basic assumption allows one to take into account the field of the valence electrons only during the SCF procedure. For this purpose, the core field is simulated by an appropriate monoelectronic potential. In this sense, the model potential framework reduces the dimension of the SCF problem without modifying the structure of the Fock operators, but changing the form of the monoelectronic hamiltonian, which is written like the exact one plus some correction terms. These correction terms include the shielding of the nucleus by the inner core electrons and the proper density distribution of these electrons.

Model potential calculations will be very useful when dealing with molecules containing higher period atoms.

The monoelectronic operator for electron i proposed by Huzinaga takes the form

$$h_i = -\frac{1}{2}\nabla_1^2 - \sum_A \{ \frac{(Z_A - X_A)}{r_{Ai}} (1 + \sum_{J_A} A_{J_A} \exp(-\alpha_{J_A} r_{Ai}^2)) + \sum_{K_A} B_{K_A} |K_A\rangle\langle K_A| \}$$

where X_A is the number of core electrons of atom A; $\{A_{J_A}\}$, $\{\alpha_{J_A}\}$ and $\{B_{K_A}\}$ are parameters which depend on the core structure of atom A, and $\{|K_A\rangle\langle K_A|\}$ are the projectors over the core orbitals of atom A.

IX. Miscellaneous Remarks

In this chapter the most important items not developed in the previous considerations of SCF theory will be given.
There is a wide variety of problems related to the main theme. We have tried to gather them in this chapter in order to obtain, as far as possible, a more complete picture of small details and important connections of the theory.

1. A Synthetic Approach

The energy expression encountered in section 1.II can be transformed into a more compact form using the equivalence of Coulomb and exchange operators, that is

$$E = \sum_i \sum_j (\omega_{ij} h + \sum_k \sum_\ell a_{ij}^{k\ell} \, (ij|k\ell)) \ .$$

In this case the related Fock operators will have the form

$$F_{ij} = \frac{1}{2}\omega_{ij} h + \sum_k \sum_\ell a_{ij}^{k\ell} \, V_{k\ell} \ ,$$

although the structure of the Euler equations and coupling operators will continue to be of the same form as in the equation of section 1.II. The reason for adopting, from the beginning, an apparently more complicated approach is simply that this is the easier way to derive the general monoconfigurational open and closed shell formulae, from equation (1.II).

2. The Concept of Shell

The SCF structure defined in section 9.II can be simplified if the total MO index set M can be partitioned into a union of disjoint subsets $\{S_I\}$

$$M = \bigcup_I S_I \ ,$$

in such a manner that the orbitals $\{|i\rangle\}$ share the same Fock operator:

$$\forall \ i \in S_I \longrightarrow F_i = F_I \ ,$$

with the Fock operator associated with the subset S_I defined as

$$F_I = \frac{1}{2}\,\omega_I h + \sum_J \sum_{j \in S_J} (\alpha_{IJ} J_j - \beta_{IJ} K_j) \;,$$

which is the same as saying that the corresponding Euler equation will be

$$\forall\, i \in S_I : \quad F_I|i\rangle = \sum_J \sum_{j \in S_J} \lambda_{ij}|j\rangle \;.$$

If it is possible to find a unitary matrix which diagonalizes the submatrix $\{\lambda_{ij}\}$; $i,j \in S_I$ leaving F_I and the energy invariant, then one has

$$\forall\, i \in S_I : \quad F_I|i\rangle = \lambda_{ii}|i\rangle$$

$$+ \sum_{J \neq I} \sum_{j \in S_J} \lambda_{IJ}|j\rangle \;,$$

thus reducing the structure of the gradient part of the coupling operator to

$$R_G = \sum_I \Pi_I \, F_I \, \Pi_I$$

and the hermitean conditions part to

$$R_H = \sum_I \sum_J \Theta_{IJ} \, P_I (F_J - F_I) \, P_J \;,$$

with $\qquad P_I = \sum_{i \in S_I} |i\rangle\langle i|$

and $\qquad \Pi_I = P_I + P_V \;.$

If these manipulations are possible, then the eigenvector subset $\{|i\rangle\}$ with $i \in S_I$ is called a shell.
The concept can be easily extended to more complicated cases.

3. Symmetry

Although there are no necessary symmetry restrictions to the development of the formalism in the preceding pages, it will be worthwhile to briefly develop the symmetry structured forms of Fock operators and coupling operators.
In fact, the coupling operator itself will be factorized as a direct sum

$$R = \bigoplus_{\alpha} R_\alpha$$

of $\{R_\alpha\}$ operators each belonging to a symmetry species. The form of each symmetry coupling operator R_α will remain the same as the non-symmetry adapted operator, but the dependence of all MO indices, which are involved in the summations, on symmetry considerations needs to be explicitly stressed. If the whole set of MO indices, M, is expressed as a union of disjoint index subsets $\{M_\alpha\}$

$$M = \bigcup_{\alpha} M_\alpha \; ,$$

each subset containing the orbitals of each symmetry species, then R_α will be defined as

$$R_\alpha = \sum_{i \varepsilon M_\alpha} \Pi_i \, F_i \, \Pi_i$$

$$+ \sum_{i,j \varepsilon M_\alpha} \Theta_{ij} \, P_i (F_j - F_i) \, P_j$$

in the framework of section 9.II, which can be easily generalized to other more complicated forms.
It is interesting to point out that the terms

$$P_i (F_j - F_i) \, P_j \text{ with } i \varepsilon M_\alpha, \; j \varepsilon M_\beta; \; \alpha \neq \beta$$

will vanish. Thus, no hermitean conditions are needed between orbitals or shells belonging to different symmetry species. The structure of the Fock operators, itself, will be simplified. So, we will have

$$F_i = \frac{1}{2} \omega_i h + \sum_{\beta} \sum_{j \varepsilon M_\beta} (a_{ij} \, J_j - b_{ij} \, K_j) \text{ with } i \varepsilon M_\alpha .$$

The energy can then be expressed as a sum of the contributions from each symmetry species:

$$E = \sum_{\alpha} \sum_{i \varepsilon M_\alpha} \langle i | F_i + \frac{1}{2} \omega_i h | i \rangle \ .$$

4. Optimization of Non-Linear Parameters

The electronic energy does not depend only on the MO's linear coefficients but also on other non-linear parameters: the AO exponents and molecular geometry. The energy functional, in this manner, can be taken as a function of the variational MO's coefficients (C) and some non-linear parameters (X). This situation can formally be expressed as E(C,X). Given a fixed set of parameters X_F, the SCF procedure is able to optimize the quantities C, and the convergence of the process will result in the calculation of an optimal set C_o. Schematically, the SCF variational process can be represented by

$$\left.\frac{\partial E(C,X)}{\partial C}\right|_{X_F} = 0 \longrightarrow E(C_o,X_F) \ .$$

If X_F is close to the optimum non-linear parameter set, one can develop $E(C_o,X)$ as a Taylor series in the neighbourhood of X_F as follows:

$$\begin{aligned} E(C_o,X) = E(C_o,X_F) &+ \\ &(X - X_F)^T \left[\left.\frac{\partial E(C_o,X)}{\partial X}\right|_{X_F} \right] + \\ &+ \frac{1}{2} (X - X_F)^T \left[\left.\frac{\partial^2 E(C_o,X)}{\partial X^2}\right|_{X_F} \right] (X - X_F) + \ldots , \end{aligned}$$

where

$$\left.\frac{\partial E(C_o,X)}{\partial X}\right|_{X_F} \quad \text{is the gradient and}$$

$$\left.\frac{\partial^2 E(C_o,X)}{\partial X^2}\right|_{X_F} \quad \text{the hessian matrix of } E(C_o,X)$$

at the point $X = X_F$. These matrices can be obtained if one is able to

retrieve sufficient analytical or numerical information of the variation of E with respect to the non-linear parameters.
An estimate of the optimum value of X will be given by

$$X_o = X_F - \frac{1}{2}\left(\frac{\partial^2 E(C_o,X)}{\partial X^2}\bigg|_{X_F}\right)^{-1}\left(\frac{\partial E(C_o,X)}{\partial X}\bigg|_{X_F}\right).$$

So, in principle, the following steps can be used in order to obtain the completely optimized energy.

a) Starting with an approximate X_F use SCF to obtain the optimum linear coefficients C_o.
b) Use $E(C_o,X)$ within a non-linear optimization process to obtain a new estimate of non-linear parameters X_o.
c) Use again $E(C,X_o)$ to obtain a refined linear set $C_o^{(1)}$.
d) Repeat the procedure until some minimal energy variation threshold is reached.

Needless to say, such a procedure, until much greater computational speed is achieved, will become unfeasible as the number of non-linear parameters increases. But, on the other hand, one should be aware of this ultimate possibility, which is already a common feature of the atomic SCF programs.

5. Generalized Brillouin's Theorem and Off-Diagonal Hermitean Conditions on Lagrange Multipliers

Brillouin's Theorem deals with the interaction of a groundstate wavefunction Φ_o and the set of monoexcited state wavefunctions $\{\Phi(i \rightarrow j)\}$, where an electron from MO $|i\rangle$ has been promoted to the MO $|j\rangle$.
Brillouin's Theorem states that if Φ_o is attached to a variational energy extremum, then

$$V(o,i \rightarrow j) = \langle\Phi_o|\mathcal{H}|\Phi(i \rightarrow j)\rangle = 0$$

for all the functions belonging to the set $\{\Phi(i \rightarrow j)\}$.
In general, the integrals $\{V(o,i \rightarrow j)\}$ take the form

$$V(o,i \rightarrow j) = \sum_p\{(a_{jp}h_{ip} - a_{ip}h_{jp}) + \sum_{k,\ell} (b_{jp}^{k\ell}(ik|p\ell) - b_{ip}^{k\ell}(jk|p\ell))\}$$

where $\{a_{pq}\}$ and $\{b_{pq}^{k\ell}\}$ are appropriate parameters depending on Φ_o and $\Phi(i \rightarrow j)$.

At the same time in a general SCF context, with Fock operators defined through

$$F_{ij} = \frac{1}{2}\,\omega_{ij}h + \sum_{k,\ell} (\alpha_{ij}^{k\ell}\, V_{k\ell} - \beta_{ij}^{k\ell}\, X_{k\ell}) \ ,$$

the hermitean conditions on Lagrange multipliers can be written as

$$\sum_p (\langle j|F_{ip}|p\rangle - \langle p|F_{pj}|i\rangle) = 0 \ ,$$

or

$$\sum_p \{\tfrac{1}{2}(\omega_{ip}h_{jp} - \omega_{pj}h_{pi}) + \sum_k \sum_\ell (\alpha_{ip}^{k\ell}(k\ell|jp) - \alpha_{pj}^{k\ell}\ (k\ell|pi))$$

$$- \sum_k \sum_\ell (\beta_{ip}^{k\ell}(kj|\ell p) - \beta_{pj}^{k\ell}(ki|\ell p))\} = 0 \ .$$

This expression can be rearranged into a structure similar to Brillouin's theorem using the symmetry properties of the state parameters and integrals, so that

$$\sum_p\{(\tfrac{1}{2}\omega_{ip}h_{jp} - \omega_{jp}h_{ip}) + \sum_k \sum_\ell ((\alpha_{ip}^{k\ell} - \beta_{ik}^{p\ell})(jp|k\ell) - (\alpha_{jp}^{k\ell} - \beta_{jk}^{p\ell})(ip|k\ell))\} = 0$$

is a generalized expression equivalent to $V(o,i \rightarrow j)$.
This result indicates the close relationship between Brillouin's theorem and the hermitean conditions of Lagrange multipliers at self-consistency. In fact, these conditions assure that Brillouin's theorem will be fulfilled. As has been stressed before in Chapter II, neglecting these conditions implies that the final result is dependent on the starting vectors in the SCF procedure.
In order to visualize these results, let's take the Li atom as an example. This is one of the simplest systems where hermitean conditions of Lagrange multipliers play an important role in obtaining correct energies. A monoconfigurational electronic energy expression for this system is given by

$$E = 2\ h_{11} + h_{22} + 2\ J_{11} - K_{11} + 2\ J_{12} - K_{12},$$

where the orbitals are $|1\rangle = (1s)$ and $|2\rangle = (2s)$. The Fock operators related to this energy are

$$F_1 = h + 2\,J_1 - K_1 + \tfrac{1}{2}(2\,J_2 - K_2)$$

and

$$F_2 = \tfrac{1}{2}\,h + \tfrac{1}{2}(2\,J_1 - K_1)\ .$$

The hermitean conditions of Lagrange multipliers take the simple form

$$\langle 2|F_1 - F_2|1\rangle = h_{12} + (11|12) + (12|22) = 0\ ,$$

and it is easy to see that this equality will not be fulfilled by just any set of vectors, but by a very particular one. If we write the groundstate configuration as $(1s)^2(2s)$, then the previous conditions are equivalent to considering the hamiltonian matrix element, between the groundstate wavefunction and the monoexcited wavefunction associated with the configuration $(1s)(2s)^2$, as null.

6. Error Analysis

In any particular case, the practical solution of the LCAO form of the Coupling Operator pseudosecular equation is associated with an error accumulation which depends on the proper Operator matrix structure and on the number of iterations used. In fact, the main source of error will be attributable to triple matrix products of the type

$$R = M^+\,F\,M\ .$$

If the matrix of errors of M is E_M and if E_F corresponds to the errors in the elements of F, the first order error matrix expression in R can be written as

$$E_R = E_M^{\ +}\,F\,M + M^+\,F\,E_M + M^+\,E_F\,M + O(2)$$

or, specifically for each element as

$$e_{R,ij} = \sum_{p,q}(e_{M,pi}\,F_{pq}\,M_{qj} + M_{pi}\,F_{pq}\,e_{M,qj} + M_{pi}\,e_{F,pq}\,M_{qj})\ .$$

Assuming that

$$e_{M,ij} \simeq e_{F,ij} \simeq \bar{e}\ ;\ \forall_{i,j}\ ,$$

then

$$e_{R,ij} \simeq \bar{e} \sum_{p,q} (F_{pq} M_{qj} + M_{pi} F_{pq} + M_{pi} M_{qj})$$

or

$$e_{R,ij} \simeq \bar{e} A_{ij} ,$$

A_{ij} being the sum of products in the previous equation. Taking a mean value of A_{ij}, say $\bar{a}$, then

$$e_R \simeq n^2 \bar{e} \bar{a}$$

where n is the number of AO used as a basis set. Assuming $\bar{a}$ is constant throughout the iterations, at the Ith iteration one can write

$$e_R^{(I)} \simeq (n^2 \bar{a}) e_R^{(I-1)} \simeq (n^2 \bar{a})^I \bar{e} .$$

7. Mathematical Structure of SCF

We will give here an alternative form of the energy functional optimization, where instead of taking the LCAO coefficients $\{C_{i\mu}\}$ as variational parameters one uses the direct product matrix elements

$$P_{i\mu\nu} = C_{i\mu} C_{i\nu}^* .$$

The starting energy functional we will use here is

$$E = \sum_i \omega_i h_{ii} + \sum_i \sum_j (\alpha_{ij} J_{ij} - \beta_{ij} K_{ij}) ,$$

which in LCAO form can be written as

$$E = \sum_i \sum_\mu \sum_\nu \omega_i P_{i\mu\nu} h_{\mu\nu}$$

$$+ \sum_i \sum_j \sum_\mu \sum_\nu \sum_\lambda \sum_\sigma (\alpha_{ij} P_{i\mu\nu} P_{j\lambda\sigma} - \beta_{ij} P_{i\mu\lambda} P_{j\nu\sigma}) (\mu\nu|\lambda\sigma) .$$

Thus, one can look at the energy functional as a quadric function of the projector elements $\{P_{i\mu\nu}\}$.

One can easily compute the gradient of this function, the elements of which will be given by

$$[\nabla(E)]_{k\alpha\beta} = \frac{\partial E}{\partial P_{k\alpha\beta}} = \omega_k h_{\alpha\beta} +$$

$$2 \sum_j \sum_\lambda \sum_\sigma \{\alpha_{kj}(\alpha\beta|\lambda\sigma) - \beta_{kj}(\alpha\lambda|\beta\sigma)\} P_{j\lambda\sigma}$$

and the hessian matrix will be simply

$$[\nabla \otimes \nabla(E)]_{k\alpha\beta;\ell\rho\tau} = \frac{\partial^2 E}{\partial P_{k\alpha\beta}\,\partial P_{\ell\rho\tau}}$$

$$= 2\{\alpha_{k\ell}(\alpha\beta|\rho\tau) - \beta_{k\ell}(\alpha\rho|\beta\tau)\}$$

$$= 2\,V_{k\alpha\beta;\ell\rho\tau}\ .$$

The extremum condition is thus:

$$\nabla(E) = 0\ ,$$

or in matrix form

$$U_{k\alpha\beta} + 2 \sum_{\ell\rho\tau} V_{k\alpha\beta,\ell\rho\tau}\, P_{\ell\rho\tau} = 0$$

using $U_{k\alpha\beta} = \omega_k h_{\alpha\beta}$.

Then, defining the corresponding matrices we will have:

$$U + 2\,VP = 0\ ,$$

so the extremum function will be obtained at

$$P = -\frac{1}{2} V^{-1} U\ ,$$

which is a well known result from studies done on the behaviour of quadric functions. The existence of V^{-1} is the main problem here. E(P) will be a minimum if $V > 0$. This is the same as saying that there exists an upper triangular matrix T such that

$$V = T^{+}T ,$$

or that V admits the Cholesky's transformation. If V^{-1} exists, there is a unique solution of the problem. If this is not so, then $V \geqslant 0$ and there appears an infinite set of solutions. The gradient condition $\nabla(E) = 0$ will no longer give a minimum, but an extremum (a saddle point). Some boundary conditions should be imposed in order to have an adequate solution to an extremum of E. In the classical SCF theory those boundary conditions are given by the orthogonality relations between MO's $\{C_i\}$ which in LCAO form are

$$C_i^{+} \; S \; C_j = \delta_{ij} .$$

Now, as one can write

$$P_i = C_i C_i^{+} ,$$

then the boundary conditions here will be

$$P_i S P_j = \delta_{ij} P_i .$$

So one should add to the energy functional some functional of the matrix

$$M = \sum_i \sum_j (P_i S P_j - \delta_{ij} P_i) .$$

The functional we seek should be a positive definite function in order to assume a minimum form when null. A norm of M, $||M||$, will suffice for our purposes. The most adequate of such norms will be the euclidean norm of M that is

$$||M||_2 = \sum_\mu \sum_\nu |M_{\mu\nu}|^2 ,$$

but this will transform the quadric simplified energy functional form into a quadric function of the projector elements. We are caught in this manner, in a vicious circle, where we always face some sort of quartic structured functional.

It will be worthwhile to stress that, if the projector conditions are not used along the optimal search, then for some optimization of the non-linear parameters of E, the functional will tend to an infinite negative sink.

X. The Problem of the Helium Atom First Excited Singlet State

1. A Possible Solution

In an open shell monoconfigurational scheme excited states with the same symmetry and spin multiplicity as groundstate cannot be properly calculated within a SCF framework, due to the nature of Rayleigh-Ritz variational principle.

The simplest case of this limitation is the helium atom first excited singlet state. The straightforward solution to the problem requires the use of a multiconfigurational framework. An alternative approach, within the monoconfigurational coupling operator formalism, is related to deflation techniques of eigenvalue problems.

Suppose that the groundstate energy (E_0) and wavefunction (Φ_0) of He are known, as well as the excited singlet energy (E_1) and wavefunction (Φ_1). It is clear that $E_o < E_1$, and if $\mathcal{H}_o$ is the hamiltonian operator that

$$\mathcal{H}_0 \Phi_I = E_I \Phi_I, \quad (I = 0,1).$$

A new operator can be constructed at this stage by means of the shift of $\mathcal{H}_0$ defined by

$$\mathcal{H}_1 = \mathcal{H}_0 - E_0 |\Phi_0\rangle\langle\Phi_0|.$$

The expected values of $\mathcal{H}_1$ will be

$$\langle \mathcal{H}_1 \rangle_I = \langle \Phi_I | \mathcal{H}_1 | \Phi_I \rangle = E_I - E_0 |\langle \Phi_0 | \Phi_I \rangle|^2, \quad (I = 0,1).$$

That is, if the wavefunctions are supposed to be normalized,

$$\langle \mathcal{H}_1 \rangle_0 = 0$$

and

$$\langle \mathcal{H}_1 \rangle_1 = E_1 - E_0 |\langle \Phi_0 | \Phi_1 \rangle|^2,$$

provided that

$$\langle \mathcal{H}_1 \rangle_1 < 0.$$

As a result, the new mathematical groundstate is the excited one. The next step consists of supposing E_0 and Φ_0 are known and constant, and applying the variational principle to the functional

$$< \mathcal{H}_1 >_1 = <\Phi_1| \mathcal{H}_0|\Phi_1> - E_0|<\Phi_0|\Phi_1>|^2 .$$

In order to obtain the variation of $< \mathcal{H}_1>_1$, let's consider $\{\phi_0\}$ as the groundstate MO and $\{\phi_1,\phi_2\}$ the excited state ones. Then, we can write

$$\Phi_0 = |\phi_0,\bar{\phi}_0|$$

and

$$\Phi_1 = (2)^{-1/2}\{|\phi_1,\bar{\phi}_2| - |\bar{\phi}_1,\phi_2|\} ,$$

so the scalar product $<\Phi_2|\Phi_1>$ can be easily evaluated in terms of involved MO scalar products

$$<i|j> = \int \phi_i^* \phi_j \, dV; \quad (i,j = 0,1,2) ,$$

so

$$<\Phi_0|\Phi_1> = (2)^{1/2}<1|0><2|0> = (2)^{1/2}<1|0><0|2> .$$

Also, we will have

$$< \mathcal{H}_1>_1 = <1|h|1> + <2|h|2> + (11|22) + (12|12)$$

$$-2 E_0|<1|0><0|2>|^2 ,$$

where h is the monoelectronic hamiltonian and (11|22), (12|12) are the bielectronic integrals involving $\{\phi_1,\phi_2\}$. Variation of $< \mathcal{H}_1>_1$ produces the Fock operators

$$F_1 = h + J_2 + K_2 - 2 E_0|<2|0>|^2 |0><0|$$

$$F_2 = h + J_1 + K_1 - 2 E_0|<1|0>|^2 |0><0| ,$$

where J_i and K_i are the usual Coulomb and exchange operators, respectively, and $|0><0|$ is a projector over the space generated by $\{\phi_0\}$. The

Euler equations will be in this case:

$$F_1|1\rangle = \lambda_{11}|1\rangle + \lambda_{12}|2\rangle$$

$$F_2|2\rangle \quad \lambda_{21}|1 \quad + \lambda_{22}|2\rangle$$

and $\lambda_{12} = \lambda_{21}^*$.

Now, calling $\alpha_i = 2\ E_0|\langle i|0\rangle|^2$,

gives

$$F_1 = F_{S2} - \alpha_2\ P_0$$

and

$$F_2 = F_{S1} - \alpha_1\ P_0$$

with obvious definitions holding for all the new symbolic operators.
A computational scheme which does not take into account the corrections of the previous discussion will deal with the Fock operators transformed into F_{S2} and F_{S1}, respectively. Such a simplification will suppose that $\langle i|0\rangle = 0$, or the groundstate MO is orthogonal to both of the excited orbitals, or simply to one of them. However, such a procedure will give an energy E_1 at SCF convergence which will tend towards the corresponding triplet energy of the system when orbital non-linear parameters are optimized. This SCF behaviour is very general and can be found in other calculations where the excited singlet has the same symmetry as the groundstate. The term "triplet catastrophe" has been coined to describe this drawback.
A consideration of the operator corrections $\alpha_i P_0$ on each Fock operator provides the SCF procedure with a correct behaviour, and overcomes the "triplet catastrophe".
A similar procedure can be envisaged for other two-electron systems such as H_2, but extension to polyelectronic cases is too cumbersome to be reasonably applied. The main difficulty arises when one seeks to compute the scalar products of determinants like

$$\langle\Phi_1|\Phi_0\rangle\ ,$$

which except for a two-electron case have a very complicated expression.

2. The "Triplet Catastrophe"

An explanation of the SCF behaviour in the cases where the excited singlet has the same symmetry as the ground state singlet can be based on the hermitean conditions of the off-diagonal Lagrange multipliers of the system.
The excited singlet energy can be written as

$$E_S = h_{11} + h_{22} + J_{12} + K_{12}$$

and the triplet energy as

$$E_T = h_{11} + h_{22} + J_{12} - K_{12} \cdot$$

In these expressions the closed shell part has been dropped in order to gain simplicity in handling the algebraic manipulations which follow. The singlet Fock operators are

$$F_1^S = h + J_2 + K_2$$

and

$$F_2^S = h + J_1 + K_1$$

and the corresponding triplet operators are

$$F_1^T = h + J_2 - K_2$$

and

$$F_2^T = h + J_1 - K_1 ,$$

which can be combined into a unique Fock operator

$$F^T = h + J_1 - K_1 + J_2 - K_2,$$

since in the triplet, both open shell MO's are elements of the same shell. In this case, there will always be two molecular orbitals which fulfill the hermitean condition of Lagrange multipliers and can be written as

$$<1|F_2^T - F_1^T|2> = <1|J_1 - K_1 - (J_2 - K_2)|2> = 0 \;;$$

also at self-consistency

$$<1|F^T|2> = <1|h + J_1 - K_1 + J_2 - K_2|2> = 0 \;,$$

which is equivalent to setting

$$<1|h|2> = 0 \;,$$

since

$$<1|J_1 - K_1|2> = 0$$

and

$$<1|J_2 - K_2|2> = 0$$

always hold. In the singlet case at self-consistency, if one has taken care that the hermitean conditions hold through the convenient operator,

$$<1|F_2^S - F_1^S|2> = 0$$

will be obtained, which is the same as

$$<1|J_1 + K_1 - (J_2 + K_2)|2> = 0 \;,$$

which is equivalent to saying

$$(11|12) - (12|22) = 0 \;;$$

but this relation is equivalent to

$$<(1|K_1 - K_2|2)> = 0$$

or, at the same time

$$<(1|J_1 - J_2|2)> = 0 \;.$$

Therefore,

$$\langle 1|K_1 - K_2|2\rangle = \langle 1|J_1 - J_2|2\rangle$$

and, after rearranging terms one obtains

$$\langle 1|(J_1 - K_1) - (J_2 - K_2)|2\rangle = 0 .$$

This is the same as saying that the hermitean conditions imposed on the singlet Lagrange multipliers will generate the same relations in the triplet case. It follows that the singlet orbitals, obtained in this way, will diagonalize the triplet Fock operators. So if α is an arbitrary real parameter we will have

$$\langle i|F_1^S|i\rangle = \alpha + \langle i|F_i^T|i\rangle ,$$

and for both indexes one arrives at:

$$K_{12} = \frac{1}{2}\alpha ,$$

that is, K_{12} in the singlet case behaves as an arbitrary parameter.
The variational procedure gives K_{12} in such a way as to obtain the minimal energy of the singlet as the energy of the parent low lying triplet, whose total wavefunction is orthogonal to the groundstate.
In cases where the groundstate has a different symmetry than the excited singlet, this orthogonality condition is automatically fulfilled, thus preventing this variational drawback. The groundstate energy shift of the preceding section can be viewed as a device that forbids the adaptation of the singlet Fock operators to those of the triplet.

3. Further Analysis of the "Triplet Catastrophe"

If one considers a two-electron system in a representation of a bidimensional space with basis functions $\{\chi_1, \chi_2\}$, the problem is simplified by the fact that the energy depends on one parameter only. The HF orbitals can be expressed as

$$|1\rangle = (\gamma\chi_1 + \chi_2)/(\gamma^2 + 1)^{1/2}$$

and

$$|2\rangle = (\chi_1 - \gamma\chi_2)/(\gamma^2 + 1)^{1/2} .$$

In this manner the singlet state energies depend on the parameter γ. The triplet energy does not depend on γ because any unitary transformation leaves it invariant. The HF orbitals of the excited singlet should fulfill, with respect to the groundstate orbital, the already discussed orthogonality relations, in order to assure proper orthogonality between ground and excited singlet wavefunctions. Therefore

$$\langle 0|1\rangle = 0 \text{ or } \langle 0|2\rangle = 0$$

should hold. Now, if we write for the groundstate HF orbital

$$|0\rangle = (\beta\chi_1 + \chi_2)/(\beta^2 + 1)^{1/2} ,$$

then

$$\langle 0|1\rangle = 0 \quad \rightarrow \quad \beta = -1/\gamma$$

and

$$\langle 0|2\rangle = 0 \quad \rightarrow \quad \beta = \gamma ,$$

where both conditions are equivalent, since γ appears in $|1\rangle$ and $|2\rangle$ in some symmetrical form. This means that, taking into account these orthogonality conditions, the minimization of the groundstate energy is conditioning the excited singlet energy. The excited singlet energy E_S can be expressed in terms of the constant (with respect to γ) triplet energy $E_T = \omega$ and the exchange integral $K_{12}(\gamma)$:

$$E_S = \omega + 2\, K_{12}(\gamma) ,$$

with

$$2\, K_{12}(\gamma) = b + (2a - b)\, f(\gamma)$$

where

$$a = (12|12)$$

$$b = (11|11) + (22|22) - 2(11|22) - 2(12|12)$$

$$f(\gamma) = \frac{\gamma^4 + \lambda(\gamma^3 - \gamma) + 1}{(\gamma^2 + 1)^2}$$

and

$$\lambda = \frac{(11|12) - (12|22)}{(11|11) + (22|22) - 2(11|22) - 2(12|12)} .$$

The isocline function with zero slope for $f(\gamma)$ is

$$g(\gamma) = 1 + \frac{2\gamma^2}{\gamma^4 - 6\gamma^2 + 1}$$

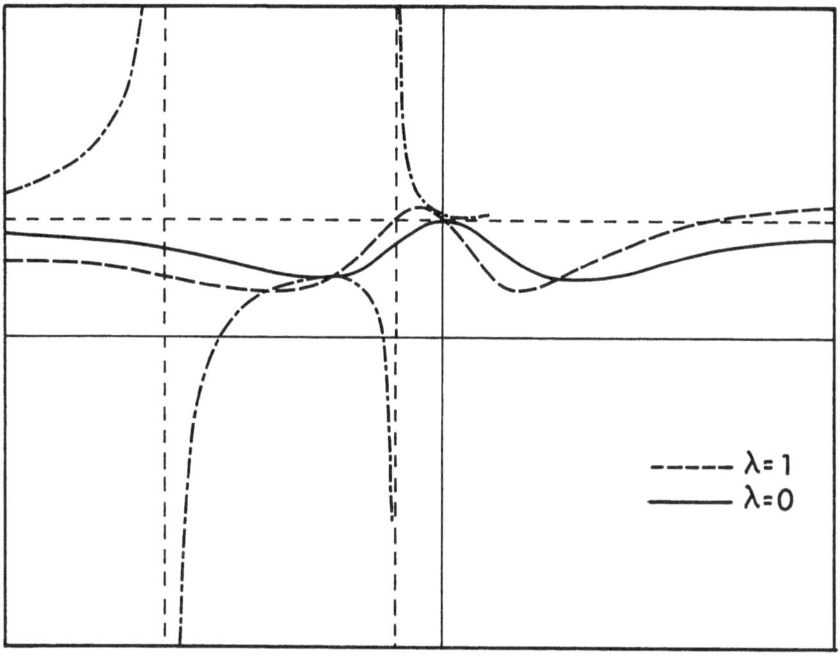

Figure 1.X. - He SCF energy functional and isoclines. See text for more details.

and is drawn in figure 1.X. In the same figure the energy curves for $\lambda = 1$ can be found. The zone of the minima is found in the open interval

$$(3 - 2\sqrt{2})^{1/2} \leqslant |\gamma| \leqslant (3 + 2\sqrt{2})^{1/2} .$$

The triplet catastrophe occurs when $K_{12} = 0$, a situation where

$$f(\gamma) = \frac{b}{b-2a} = \frac{\alpha - 1}{\alpha - 2}$$

with

$$\alpha = \frac{1}{2}\{(11|11) + (22|22) - (11|22)\}/(12|12) ,$$

which, from the integral values, always fulfills $\alpha \gtrsim 0$.
Since $K_{12} \geqslant 0$ always holds we will have:

$$f(\gamma) \geqslant \frac{\alpha - 1}{\alpha - 2} .$$

In the triplet catastrophe only the equal sign holds, and one has from the isocline equation:

$$f(\gamma) = g(\gamma_o) = \frac{\alpha - 1}{\alpha - 2} ,$$

which permits one to find the relation

$$\alpha = \frac{1}{2}(\gamma_o^2 + \gamma_o^{-2}) - 1 ,$$

where γ_o is obtained through the triplet catastrophe condition on the exchange integral. This gives the result

$$\gamma_o - \gamma_o^{-1} = 4\alpha\ \lambda(\alpha - 2)^{-1} ,$$

and using the former equation relating α and γ_o, one arrives at

$$\lambda^2 = (\alpha - 2)^2\ (8\alpha)^{-1}$$

or in integral terms

$$\frac{[(11|12) - (12|22)]^2}{[(11|11) + (22|22) - 2\{(11|22) + (12|12)\}]^2} =$$

$$\frac{[(11|11) + (22|22) - 2\{(11|22) + 2(12|12)\}]^2}{8(12|12)[(11|11) + (22|22) - 2(11|22)]} .$$

The last relation can only be obtained through exponent variation. In a more complicated electronic system, the situation is not too simple because the energy depends on more than one parameter. However, one can make some tentative proposal of the general trend by considering that only the singly occupied orbitals play a leading role in other triplet catastrophe situations.

4. Some Results on Monoconfigurational He SCF

The application of the previous discussion on the ground and excited states of He is shown in Table 1.X. A basis set of 3 STO's, two of 1s type and one 2s, has been chosen. The exponents used in each calculation are at the head of each column, and below the triplet and singlet energies are written the transition values. The second and third columns give the singlet and triplet optimal values, respectively. The fourth column gives the Hartree-Fock limit energies and the last one the experimental values.
The effect of the exponent variation is most effective in the triplet. The correlation energy is greater in the groundstate and minimal for the triplet. The transition energies are in defect when compared to the experimental ones but very similar to the Hartree-Fock limiting values.

Table 1.X

Helium $^1S_0$, $^3S_1$ and $^1S_1$ energies (a.u.)

	Basis set	(1)	(2)	(3)	HF(4)	Experiment
exponents	1s	2.0	2.0	2.002		
	1s'	0.8	0.534	0.83		
	2s	0.575	0.509	0.627		
$^1S_0$	energy	-2.8583			-2.8617	-2.9037
$^3S_1$	energy	-2.1733		-2.1742	-2.1743	-2.1752
	ΔE	0.6850			0.6874	0.7285
$^1S_1$	energy	-2.1294	-2.1330		-2.1435	-2.1460
	ΔE	0.7289			0.7182	0.7577

(1) Optimized through $^1S_0$. (2) Optimized through $^1S_1$. (3) Optimized through $^3S_1$. (4) Ch. Froese-Fischer, J. Chem. Phys., 47, 4010 (1967).

XI. Applications

1. Introduction

We pretend here to give some results of the application of the theoretical framework to some straightforward cases. In this sense we present "*ab initio*" calculations on various molecular systems, as well as some examples of empirical calculations. A picture of the practical usefulness of the discussed formalism can be easily drawn from the following results which, with sufficient information, provide for the implementation and checking of future programs.

2. SCF Study of Water: Ground and Excited States

a) Properties

The formalism described above has been used to study the water molecule in various states: ground ($^1A_1$) and excited singlets ($^1B_1$), triplet ($^3B_1$) and positive ion doublet ($^2B_1$) and quadruplet ($^4B_1$).
The coordinates have been taken for all states at the ground state equilibrium geometry, and are given in Table 1.XI. A GTO basis set of double

Table 1.XI

H_2O (a.u.) (*)

	X	Y	Z
O	0	0	0
H_1	0	1.430456	1.107118
H_2	0	-1.430456	1.107118

(*) L.C. Snyder, H. Basch, "Molecular Wave Functions and Properties", J. Wiley, New York (1972).

zeta quality has been chosen, with diffuse s orbitals added to oxygen. The results on molecular properties are shown in Table 2.XI, which gives the total energy, dipole moment, mean square distance of the electrons to the center of mass, potential and density at the nucleus, Mulliken gross atomic populations and bond orders for the O-H bond, as well as the vertical transitions between all the states studied.
The main features associated with the variation of molecular properties

upon excitation are clearly visible in the table and can be summed up as follows:

1) Dipole moment. There is a sharp decrease in dipole moment magnitude in both excited triplet and singlet, although it is stronger in the triplet.
2) $\langle r^2 \rangle$. The mean square distance increases from groundstate to the triplet and takes the largest value in the excited singlet. This can be associated with the Rydberg nature of these states. For the positive ion, this quantity decreases with respect to the groundstate due to the loss of one electron, and again increases in the quadruplet, which is an excited state of the ion.
3) $\langle 1/r \rangle$ and $\langle \delta \rangle$. Potential and density at the nucleus both behave in a similar manner. The oxygen contribution remain practically invariant while small changes occur in the hydrogen part, since the potential on hydrogen is affected most.
4) Gross atomic populations and bond orders. An electronic transfer from O to H takes place in the triplet and singlet excitations. This is connected to the variation of dipole moment, as the O-H bond orders decrease a considerable amount, and are negative in the ions. The ionization affects the charges on hydrogens in the doublet, giving practically a neutral oxygen, but the pattern changes in the quadruplet giving a reversed situation.

Some of the spectacular changes in molecular behaviour can be seen in an adequate way through figures 1.XI to 8.XI, where the electrostatic potential maps and density functions for various states are shown.

Figures 1.XI, 2.XI, and 3.XI show the electrostatic molecular potential calculated in the molecular plane (YZ) for the ground, triplet and excited singlet, respectively. The high ridge at the center of the square corresponds to the repulsive region which surrounds the molecular frame, the hydrogen atoms being directed to the left of the figures. In the groundstate a deep potential well appears in the direction of the negative Z-axis as shown by the funnel shaped surface at the right of figure 1.XI. This feature disappears in the triplet and excited singlet (figures 2.XI and 3.XI, respectively) and is replaced by a shallow minimum. Both triplet and excited singlet surfaces have practically the same shape. It is interesting to note that in these states the minimum site moves to the region of the oxygen lone pair center of charge.

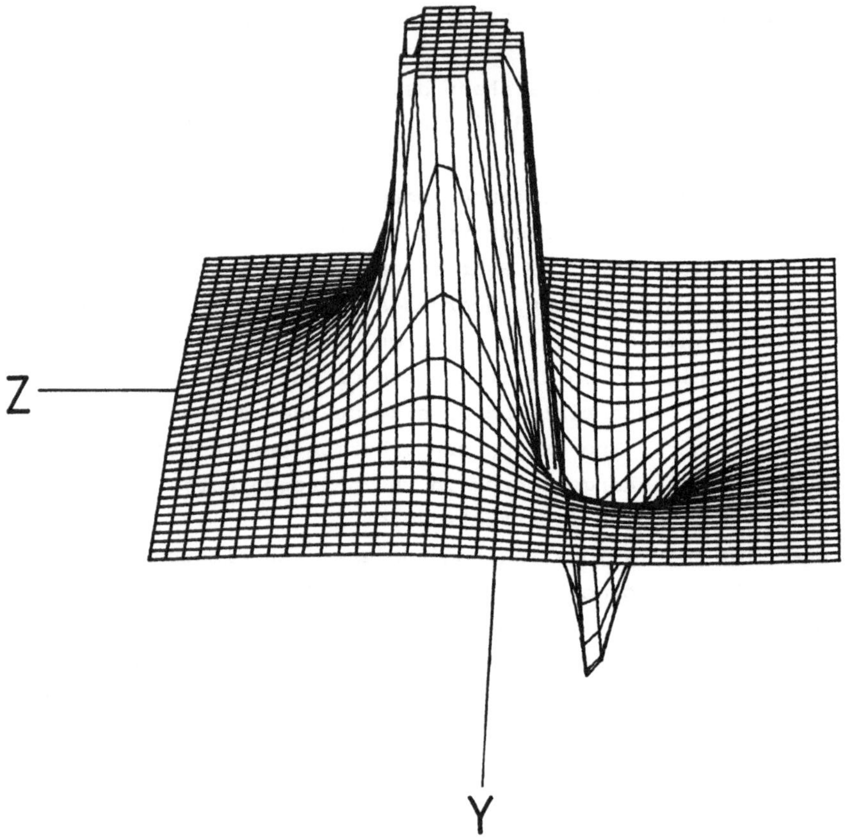

Figure 1.XI. - H_2O electrostatic molecular potential ground state ($^1A_1$) plane YZ

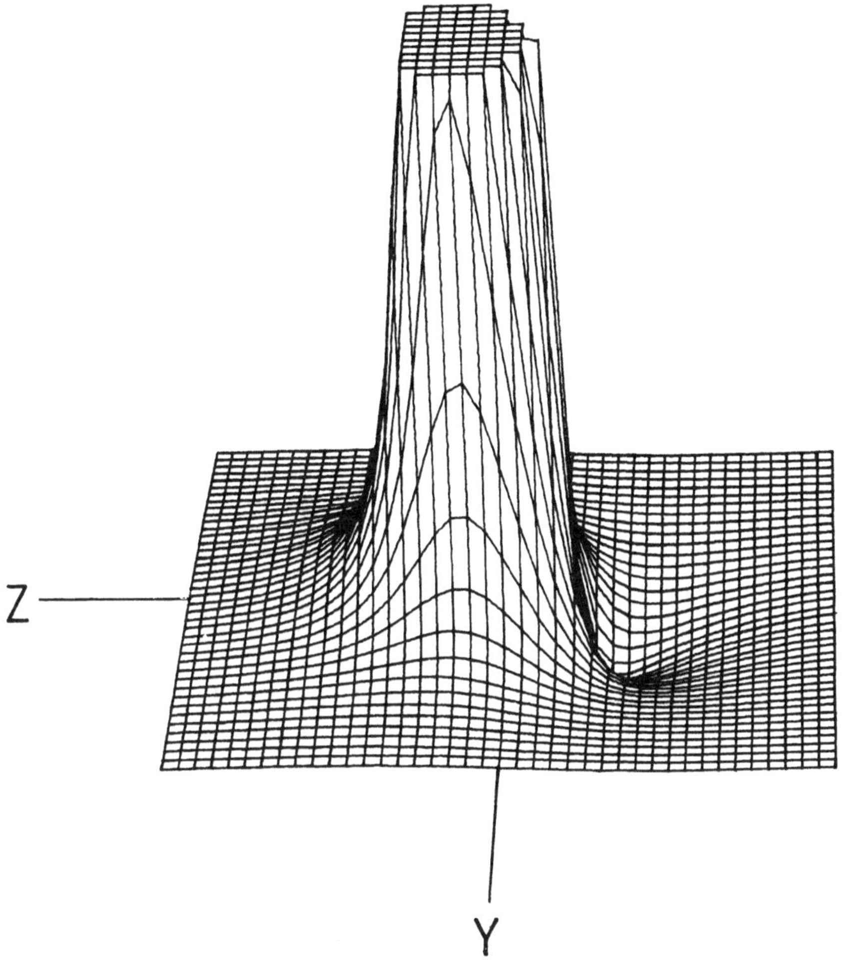

Figure 2.XI. - H_2O electrostatic potential triplet state ($^3B_1$) plane YZ

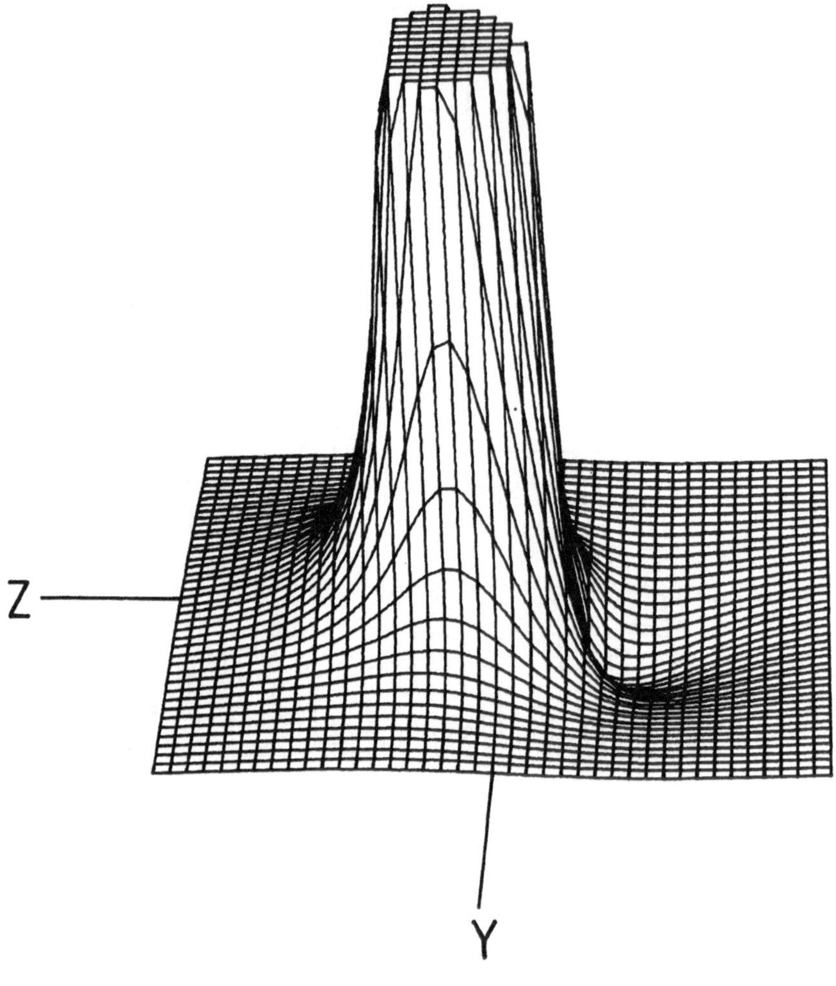

Figure 3.XI. - H_2O electrostatic potential excited singlet state ($^1B_1$) plane YZ

The displaced minimum in the triplet and excited singlet is in the plane XZ, which bisects the HOH angle. This is shown in figure 4.XI, where the electrostatic potential surface of the triplet is drawn. The main

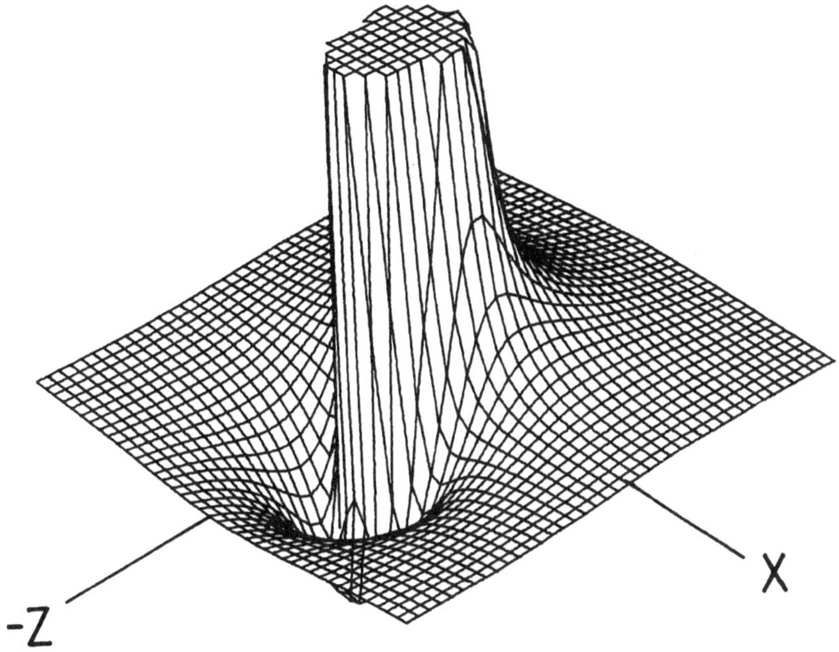

Figure 4.XI. - H_2O electrostatic potential triplet state ($^3B_1$) plane XZ

conclusion which one can obtain from these figures affects the shape of the H_3O^+ ion. In the state $^1A_1$ the ion will be planar but in the $^3B_1$ and $^1B_1$ excited states the ion will have a pyramidal structure. Figures 5.XI and 6.XI show the valence electron density maps for the ground and triplet states in the molecular plane. In both figures the hydrogen atoms are located on two small hills at the front and left corners, while the mountain at the center of the figure is the oxygen contribution to the valence part. Once again, the negative Z-direction in the ground state is asymmetric, while both Z directions are very similar in the triplet. In both states there is some valley surrounding the central atom. A planar view of the density contours in the plane XZ for states $^1A_1$ and $^3B_1$ is shown in figures 7.XI and 8.XI. The ground state electronic cloud is polarized in the X direction as a consequence of having

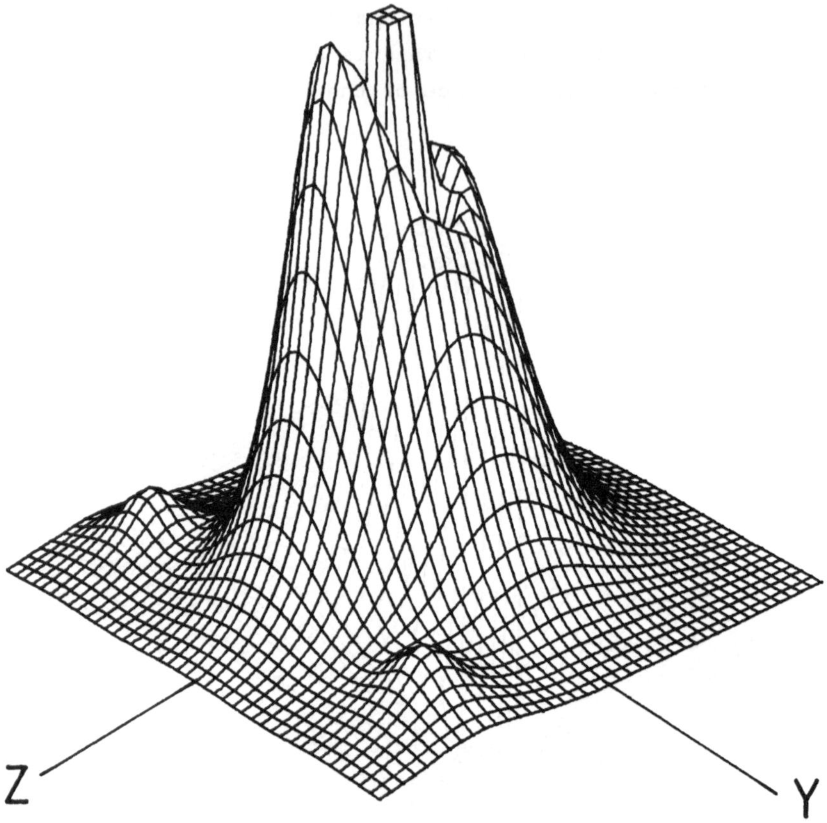

Figure 5.XI. - H_2O valence electron density function. Groundstate ($^1A_1$) plane YZ

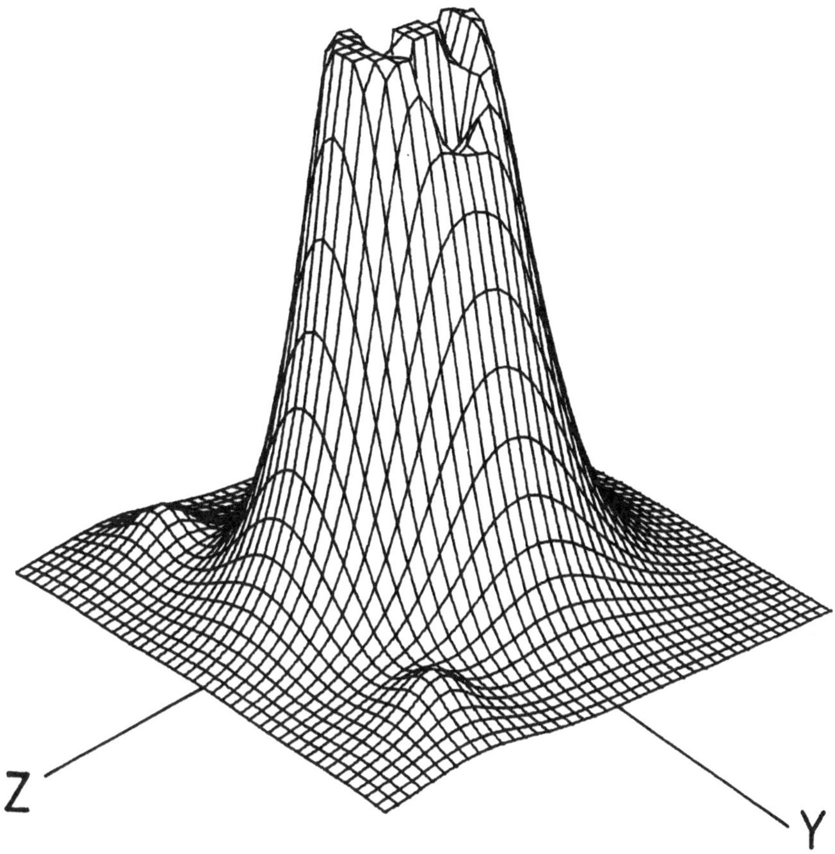

Figure 6.XI. - H_2O valence electron density function. Triplet state ($^1B_1$) plane YZ

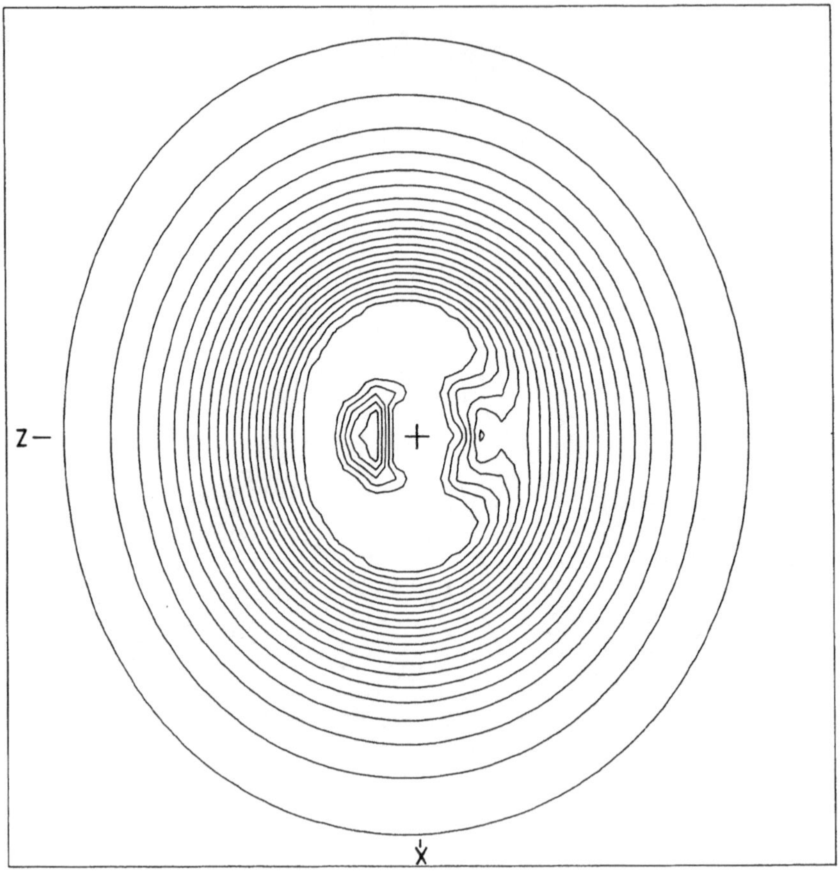

<u>Figure 7.XI</u>. - H_2O valence electron density function. Groundstate ($^1A_1$) plane XZ

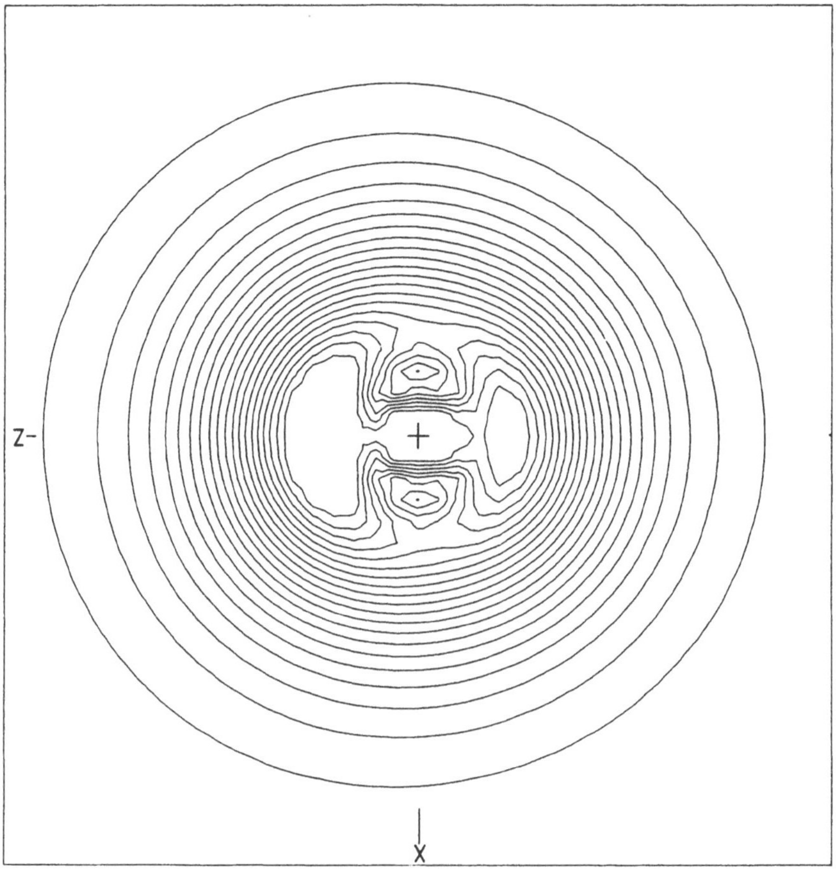

Figure 8.XI. - H_2O valence electron density function. Triplet state ($^3B_1$) plane XZ

two electrons in a π-type orbital. The excitation giving the $^3B_1$ state is of π-σ* type, and results in a polarization of the density cloud towards the Z axis. It follows that directionally different chemical behaviour is to be expected when considering ground or excited states of H_2O.

Table 2.XI

H_2O molecular properties[a] (a.u.)

State	Energy	μ	$<r^2>$
$^1A_1$	-76.001453	1.02	19.26
$^3B_1$	-75.772771	0.47	32.11
$^1B_1$	-75.755966	0.60	34.34
$^2B_1$	-75.592122	-	14.43
$^4B_1$	-15.146091	-	21.09

State	$<1/r>$		$<\delta>$		Q		P(O-H)
	O	H	O	H	O	H	
$^1A_1$	-22.34	-0.9504	295.14	0.3587	8.6739	0.6630	0.5682
$^3B_1$	-22.13	-0.7865	296.46	0.3170	8.4407	0.7797	0.1926
$^1B_1$	-22.11	-0.7614	296.45	0.3140	8.5761	0.7119	0.1097
$^2B_1$	-21.81	-0.4680	295.84	0.3159	7.9600	0.5200	-0.5300
$^4B_1$	-21.51	-0.3729	296.85	0.2682	7.3025	0.8488	-0.5200

Vertical transition energies

	$^3B_1$	$^1B_1$	$^2B_1$	$^4B_1$
$^1A_1$	0.229	0.245	0.409	0.855
$^3B_1$	-	0.017	0.181	0.627
$^1B_1$	-	-	0.164	0.610
$^2B_1$	-	-	-	0.446

[a] Basis set from: Th. H. Dunning, P.J. Hay; Modern Theoret. Chemistry Volume 2. H.F. Schaeffer Ed., Plenum Press (1977).

b) Extended Basis Set Calculation

An accurate calculation has been done for groundstate ($^1A_1$) triplet ($^3B_1$) and excited singlet ($^1B_1$) states of water. The basis set is of double zeta accuracy plus d polarization function on oxygen.
The same properties as in the previous calculation are given here in Table 3.XI. The groundstate energy is ameliorated with respect to the double zeta basis set, but the excited states are poorly described due to the lack of diffuse s orbitals on oxygen. This behaviour confirms the Rydberg nature of the excited states of water.
The inadequacy of the basis set for the excited states in this case resides mainly in the Mulliken populations, dipole moment and quadratic distance of the electrons to the center of mass.
Another calculation, whose results are shown in Table 4.XI, has been performed including polarization functions on both O and H, and diffuse orbitals on O. The energies of the three states appear highly optimized, and other changes occur in the description of molecular parameters. The diffuse s orbitals on O seem to enhance, again, the Rydberg nature of the excited states. The groundstate dipole moment is closer to the experimental value (0.728 a.u.). Small changes occur in $\langle r^{-1} \rangle$ and $\langle \delta \rangle$, and population values are completely different as expected.
In the light of these results, one should be very careful in choosing an adequate basis set for excited states. That a groundstate basis is very good does not mean that it serves to deal with excited state energies and properties. In cases like H_2O, a diffuse s orbital on atoms other than H gives a better description of the excited state than a polarization function set of d orbitals. This will always hold in the cases where the excited states have a strong Rydberg nature.

c) Geometry Optimization

The geometries of the ground and excited states had been optimized with respect of the OH distance and HOH angle of H_2O. Table 5.XI gives the calculated values using a double zeta quality basis set.
The main trend in all the excited states goes towards increasing the OH distance and HOH angle. The largest distance and angle has been found in the excited singlet, followed by the triplet and on a small scale in the doublet. Groundstate angle is overestimated in this kind of calculation, the error being of the order of ~+5°. The distance is also overestimated by about +0.1 a.u. Assuming the same error trend in the other states, one can obtain from Table 5.XI a corrected picture of the molecular shapes in the excited states.

Table 3.XI

H_2O accurate calculation[a]: Properties (a.u.)[b]

State	Energy	μ	$<r^2>$
$^1A_1$	-76.0229	1.051	19.0735
$^3B_1$	-75.7415	1.935	24.2571
$^1B_1$	-75.7111	1.937	24.7470

State	$<r^{-1}>$		$<\delta>$		Q		P(O-H)
	O	H	O	H	O	H	
$^1A_1$	-22.340	-0.9852	287.38	0.3916	8.9170	0.5422	0.5679
$^3B_1$	-22.214	-0.9122	288.94	0.3540	8.0089	0.9955	-1.5856
$^1B_1$	-22.216	-0.9081	289.04	0.3467	8.0601	0.9699	-1.7082

Vertical transition energies

	$^3B_1$	$^1B_1$
$^1A_1$	0.28	0.31
$^3B_1$	-	0.03

[a] Basis set: 1) double zeta: L.C. Snyder, H. Basch; "Molecules Wave-functions and Properties", J. Wiley, New York (1972).
2) d polarization: T.H. Dunning; J. Chem. Phys., 55, 3958 (1971).

[b] From: R. Gallifa; Tesis Doctoral, Instituto Quimico de Sarria, (1976).

Table 4.XI

H_2O extended basis set[a]: Properties (a.u.)

State	Energy	μ	$\langle r^2 \rangle$
$^1A_1$	-76.052164	0.8973	19.4175
$^3B_1$	-75.825184	1.621	55.4985
$^1B_1$	-75.812336	1.708	59.7759

State	$\langle r^{-1} \rangle$		$\langle \delta \rangle$		Q		P(O-H)
	O	H	O	H	O	H	
$^1A_1$	-22.332	-0.9883	294.8	0.4290	8.7845	0.6077	0.6584
$^3B_1$	-21.470	-0.4372	296.5	0.2864	8.9668	0.4834	0.4259
$^1B_1$	-21.415	-0.3985	296.3	0.2822	9.0511	0.4744	0.4660

Vertical transition energies

	$^3B_1$	$^1B_1$
$^1A_1$	0.2270	0.2398
$^3B_1$	-	0.0128

[a] Basis set: See Table 2.XI.

Polarization: T.H. Dunning; J. Chem. Phys., 55, 3958 (1971).

Table 5.XI

H_2O optimum geometries for various states[a]

State	r(O-H)a.u.	angle HOH°
$^1A_1$	1.942	109.3
$^3B_1$	2.119	118.2
$^1B_1$	2.121	121.1
$^2B_1$	1.990	114.0

[a] Basis set:L.C. Snyder, H. Basch; "Molecular Wavefunctions and Properties", J. Wiley, New York (1972).

3. Paired Excitation Calculations on Water

The double zeta quality basis set with diffuse s orbitals of Table 2.XI has been used to perform three MC PE SCF calculations on H_2O. The structure of the fourteen MO's of water for groundstate in terms of the chosen basis set is

MO number	1	2	3	4	5	6	7
Symmetry	$(1a_1)^2$	$(2a_1)^2$	$(1b_1)^2$	$(3a_1)^2$	$(1b_2)^2$	$(4a_1)^0$	$(2b_1)^0$

MO number	8	9	10	11	12	13	14
Symmetry	$(5a_1)^0$	$(3b_1)^0$	$(2b_2)^0$	$(6a_1)^0$	$(7a_1)^0$	$(4b_1)^0$	$(8a_1)^0$

The three calculations we have performed are:

A) 10 Configurations involving MO's {4,5,6,7,8}

B) 20 Configurations involving MO's {3,4,5,6,7,8}

C) 56 Configurations involving MO's {3,4,5,6,7,8,9,10}.

The number of configurations in each case corresponds to all the possible paired excitations which can be performed with the orbitals and electrons available.

Table 6.XI

Comparison of molecular properties (a.u.) of H_2O for each multiconfigurational scheme

Property		Number of configurations			
		1	10	20	56
Energy:	(1)	-76.001453	-76.002206	-76.004044	-76.017510
	(2)	-	-74.830451	-74.832400	-74.834000
μ		1.019	1.016	1.012	1.009
$<r^2>$		19.26	19.29	19.33	19.39
$<r^{-1}>$					
	O	-22.34	-22.34	-22.34	-22.33
	H	-0.950	-0.950	-0.949	-0.948
$<\delta>$					
	O	295.14	295.15	295.15	295.17
	H	0.3587	0.3587	0.3584	0.3589
Q					
	O	8.674	8.672	8.668	8.663
	H	0.663	0.663	0.666	0.668
P(O-H)		0.568	0.563	0.560	0.555

(1) Ground state

(2) First excited state

The results are assembled in Table 6.XI. From the table one can see the small effect on all the molecular variables produced by the multi-configuration process. This is due to the dominant role of the ground-state configuration in all calculations.
In all three cases, the first excited configurations mainly correspond to the paired excitation $1b_2 \rightarrow 4a_1$, whose energy is given also in the table below the groundstate energy.
From the technical point of view, this is a good test case for the use of a large number of open shell operators. Case A corresponds to a 6 shell case, cases B and C corresponds to 7 and 9 shells coupling operators.

4. Formaldehyde

a) Minimal Basis Set Calculation

A STO-G minimal basis set has been used here, taking 4GTO's for inner core 1s orbitals on O and C and 3GTO's for the remaining atomic orbitals. The scale factors for each orbital have been optimized, in order to give as far as possible a coherent picture of ground and excited states. The geometry is given in Table 7.XI.

Table 7.XI

Formaldehyde geometry (a.u.)[a]

	X	Y	Z
H_1	O.	1.79338799	-1.1097769
H_2	O.	-1.79338799	-1.1097769
C	O.	O.	O.
O	O.	O.	2.2825

[a] L.C. Snyder, H. Basch; "Molecular Properties and Wavefunctions", J. Wiley, New York (1972).

The ground ($^1A_1$) and excited ($^1A_2$) singlets as well as the triplet ($^3A_2$) states have been calculated. Table 8.XI gives the calculated properties. Dipole moments have been found to decrease with the excitation, but on the contrary to the H_2O molecule, the values of $\langle r^2 \rangle$ are practically invariant. The valence nature of the excited states is thus stressed. The properties for both excited states $^1A_2$ and $^3A_2$ are practically the same. The trends in both tables are similar, although the change of basis affects the values of the calculated molecular parameters. The dipole moment in groundstate is highly affected but the relative variation of this quantity behaves identically in both basis sets.
The values of potential remain unaltered but the densities on oxygen and carbon show great differences. As we have discussed in section 1.3.VIII, the gross atomic populations will change due to explicit dependence on basis set changes. Despite these differences, the relative changes due to excitation remain in the same direction.

Table 8.XI

Formaldehyde monoconfigurational SCF properties (a.u.)[a]

State	Energy	μ	$<r^2>$
$^1A_1$	-113.0725	0.612	58.6
$^3A_2$	-112.9934	0.323	58.4
$^1A_2$	-112.9738	0.419	58.4

State	$<r^2>$			$<\delta>$			Q		
	H	C	O	H	C	O	H	C	O
$^1A_2$	-1.130	-14.59	-22.14	0.371	91.54	227.97	0.939	5.923	8.200
$^3A_2$	-1.123	-14.68	-22.08	0.351	91.36	228.02	0.913	6.091	8.084
$^1A_2$	-1.118	-14.67	-22.12	0.351	91.40	228.02	0.912	6.054	8.122

Vertical transition energies

	$^3A_2$	$^1A_2$
$^1A_1$	0.0791	0.0987
$^3A_2$	-	0.0196

[a] Minimal basis set exponents used:

	H	C	O
1s	1.252	5.677	7.662
2s	-	1.762	2.245
2p	-	1.691	2.237

From: R. Gallifa; Tesis Doctoral, Instituto Quimica de Sarria (1976).

The transition energies are less than the experimental ones ($^1A_1 \rightarrow {}^3A_2$: 0.115 a.u.; $^1A_1 \rightarrow {}^1A_2$: 0.128 a.u.) due to correlation effects. The singlet-triplet splitting is greater than the experimental ($^3A_2 \rightarrow {}^1A_2$: 0.013 a.u.) due to the singlet correlation energy which is bigger than in triplet.

b) <u>Double Zeta Calculation</u>

A double zeta calculation has been performed with the same geometry as in the minimal basis set used on the three states, adding the ion doublet $^2A_2$. Table 9.XI gives the calculated properties.

Table 9.XI

Formaldehyde monoconfigurational SCF properties[a] (a.u.)

State	Energy	μ	$<r^2>$
$^1A_1$	-113.81867	1.186	60.9
$^3A_2$	-113.73192	0.595	61.3
$^1A_2$	-113.71788	0.664	61.3
$^2A_2$	-113.46584	-	52.2

State	$<r^{-1}>$			$<\delta>$			Q		
	H	O	C	H	O	C	H	O	C
$^1A_1$	-1.037	-22.338	-14.610	0.368	296.14	119.84	0.881	8.323	5.914
$^2A_2$	-1.024	-22.298	-14.691	0.339	296.33	119.65	0.808	8.076	6.307
$^1A_2$	-1.025	-22.309	-14.686	0.342	296.31	119.67	0.815	8.111	6.258
$^2B_2$	-0.656	-21.847	-14.255	0.316	296.72	119.99	0.652	7.839	5.855

Vertical transition energies

	$^3A_2$	$^1A_2$	$^2A_2$
$^1A_1$	0.0768	0.0868	0.353
$^3A_2$	-	0.0140	0.266
$^1A_2$	-	-	0.252

[a] Double zeta basis set: Th. H. Dunning, P.J. Hay; Modern Theoret. Chemistry, Volume 2, H.F. Schaeffer, Ed., Plenum Press (1977).

From the electrostatic molecular potential point of view, the H_2CO molecule has a different behaviour, very similar to the situation found also in H_2O, in front of an approaching point charge when in ground or excited state. The ground state forms a planar complex with the proton placed along the CO axis. Figure 9.XI gives a picture of the potential in the plane XZ, where the molecule is placed. The top of the ridge shows the molecular shape with the oxygen directed towards the left.

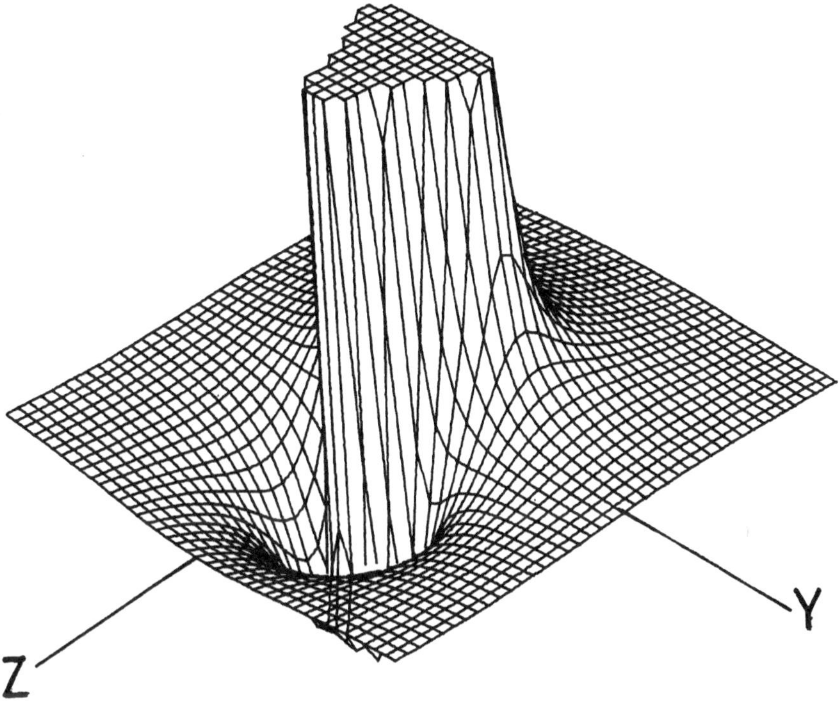

Figure 9.XI. - Electrostatic molecular potential of formaldehyde. Groundstate ($^1A_1$) plane YZ

The potential well is apparent. On the other hand, the triplet potential in the same plane, as shown in Figure 10.XI, presents a similar shape but with a shallow and narrow well. In the YZ plane, both ground (Figure 11.XI) and triplet (Figure 12.XI) give approximately the same picture. The excitation thus produces a migration of the potential minimum from the CO axis to the center of charge of the oxygen lone pairs. This is a typical feature already encountered in H_2O, and should be associated, apparently, with the oxygen behaviour.

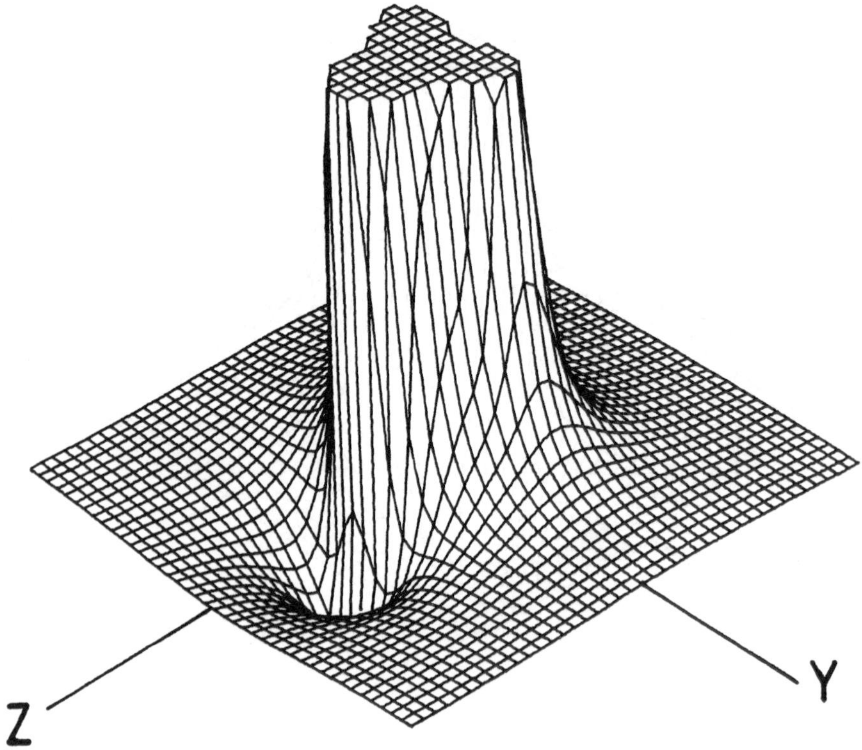

Figure 10.XI. - Electrostatic molecular potential of formaldehyde. Triplet state ($^3A_2$) plane YZ

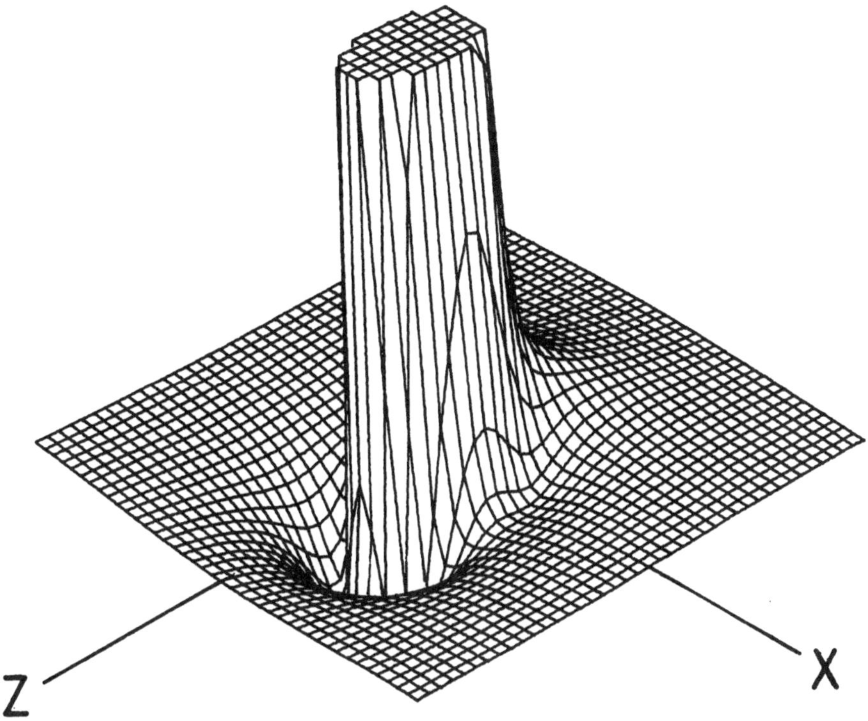

Figure 11.XI. - Electrostatic molecular potential of formaldehyde.
Groundstate ($^1A_1$) plane XZ

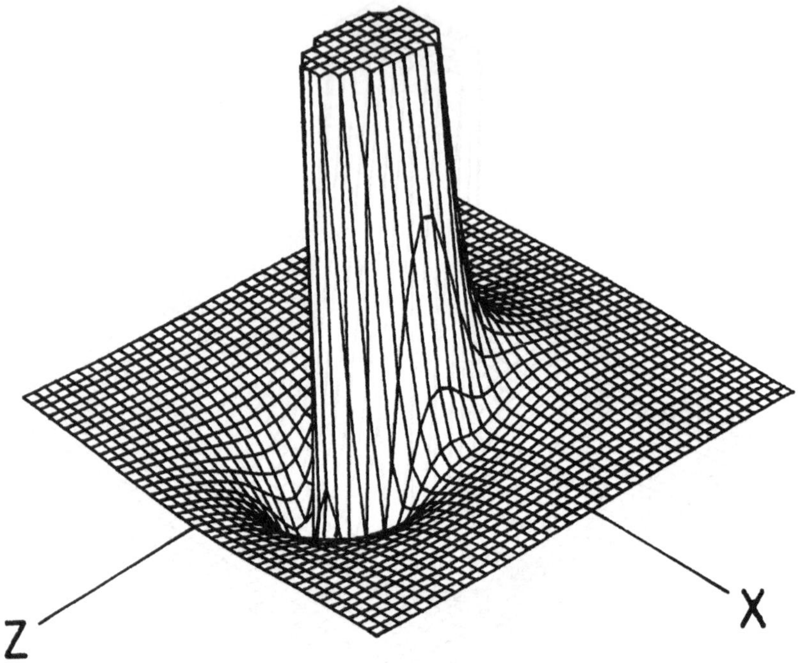

Figure 12.XI. - Electrostatic molecular potential of formaldehyde. Triplet state ($^3A_2$) plane XZ

5. Magnesium Oxyde

We have chosen this molecule for its SCF convergence peculiarities, reported by Huron and Rancurel in 1972. Apart from the wrong description of the spectrum given by the SCF calculations, the MgO problem is a good test case of the optional use of shift operators and paired excitation formalism.
The monoconfigurational calculation of $X^1\ \Sigma^+$ groundstate has a typical nonconvergent behaviour, giving energies which oscillate between two values. The shift operator prevents this anomalous behaviour produced by the interchange of the 6σ and 7σ orbitals between each iteration. The energy obtained with a double-zeta quality basis set for both atoms is -274.30267 a.u., for an interatomic distance of 3.305 a.u.
The leading role of the groundstate occupied 6σ orbital and the first virtual 7σ can be easily seen when the molecular triplet state is calculated, whose energy is -274.34941 a.u. for the same geometry. This result is a consequence of using the wrong monoconfigurational SCF structure for this molecule, the experimental groundstate being the closed shell singlet.
A paired excitation multiconfigurational SCF calculation involving the $(6\sigma)^2\ (7\sigma)^0$ and the $(6\sigma)^o\ (7\sigma)^2$ configurations gives an energy for the groundstate of -274.36615 a.u., but the configuration coefficients are -0.8801 and 0.4748 for both configurations,respectively. It follows that the 7σ orbital has a stabilizing role in the description of the $X^1\Sigma$ groundstate . A 20 configuration calculation has been performed also, involving the orbitals $(2\pi)\ (6\sigma)\ (7\sigma)\ (3\pi)$, giving an energy of -274.3844 a.u., with ground state contributions to the wavefunction very similar to those found in the biconfigurational case.
In any of the calculations, the apparent structure according to Mulliken population analysis is close to Mg^+O^-.

6. Nitrogen Dioxyde

The NO_2 molecule has been chosen because one can study, as a result, a groundstate open shell doublet with a great variety of doublet excited states.
Figure 13.XI shows the transition energies of the different states studied under different methods and basis sets. The left part of the figure corresponds to the present monoconfigurational formalism, and the right part of the figure shows a configuration interaction calculation of single excitations from the groundstate . The column named

Gangi et al. corresponds to a refinement of the SCF solution for each state through a C.I. procedure.

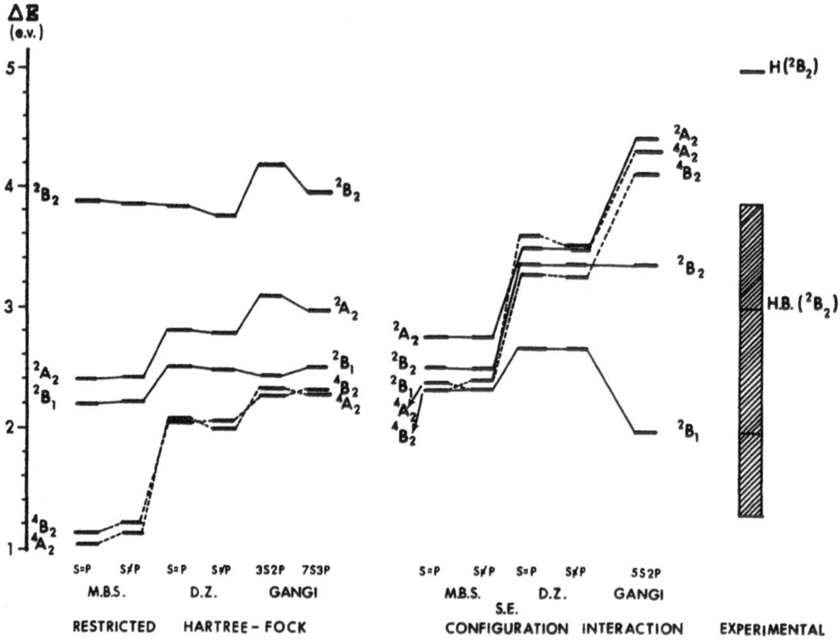

Figure 13.XI. - Energy levels of NO_2. Various basis sets are compared in SCF and C.I. frameworks. The column named Gangi displays the results of

a) L. Burnelle, A.M. May, R.A. Gangi; J. Chem. Phys., 49, 561 (1968),
b) R.A. Gangi, L. Burnelle; J. Chem. Phys., 55, 843 (1971),
c) Ibid; J. Chem. Phys., 55, 851 (1971)

The doublet $^2A_2$ and $^2B_1$ transition energies are insensitive to the basis set. The transition energy increases with the size of basis set, since correlation energies decrease to a greater extent in the groundstate than in excited states.

The SCF and C.I. results are very different for the doublet $^2B_2$. This state is poorly described in SCF, because it strongly interacts with another monoexcitation from ground state ([$4b_2$ - $6a_1$]: 0.9267 and [$1a_2$ - $2b_1$]: 0.3524). The transition energy between quadruplets is much less than expected due to the large correlation energy differences between the states involved in the transition.

7. Methanol

A minimal basis set computation of methanol has been carried out in order to have some information about the possible conformational analysis of excited states. Table 10.XI gives some of the values found for this molecule in the ground ($^1A'$) and excited ($^1A''$) singlets and for the triplet ($^3A''$).
Dipole moments do not show any big variation except for the staggered excited singlet. The value of the mean square distance increases as in the H_2O molecule, but in a less spectacular manner. It is probable that the effect will be enhanced by adding diffuse 3s orbitals to the C and O atoms. The rotation barriers decrease with excitation, a very interesting trend which should be further investigated.

Table 10.XI

Methanol[b]

State		$^1A'$	$^3A''$	$^1A''$	
Energy[a]	staggered	-114.25986	-113.90268	-113.84486	
	eclipsed	-114.25750	-113.90180	-113.84406	
μ^a	staggered	0.5977	0.5945	0.7022	
	eclipsed	0.6269	0.5383	0.6216	
$\langle r^2 \rangle^a$	staggered	82.4531	83.2722	82.9345	
	eclipsed	82.4424	83.3106	83.0213	
Rotation barrier*		1.42	0.552	0.502	(1.06)°
Transition energy+			9.72	11.29	(10.86)[x]

a) a.u.; *) kcal/mol; +) e.V.; °) Experimental value for groundstate ;

x) Ionization potential; b) From: R. Gallifa; Tesis Doctoral, Instituto Quimico de Sarria (1976).

8. Diimine

Figure 14.XI shows the degrees of freedom and symmetry characteristics of studied conformations. The twist angle of the N-N double bond is taken as a master variable; in each point of the potential surface all geometrical parameters have been optimized, however two constraints are considered in any case: both N-H distances and NNH angles are maintained equal. These constraints are needed in order to maintain a two-fold symmetry axis throughout all calculations. This symmetry element permits one to apply and discuss Woodward-Hoffmann rules, along the reaction path.

Figure 15.XI shows the correlation of the diimine M.O. (3 last occupied and the first virtual) throughout the transformation from the trans to cis forms, maintaining the symmetry axis. It should be noted that symmetry species are conserved throughout the evolution of the diagram, but not the symmetry subspecies which correspond to the most symmetrical conformation. For the sake of completeness, the organic chemical notation is also given, although it loses any meaning in the intermediate zones. The orbitals n+ and n- correspond respectively to the symmetric and antisymmetric linear combinations of lone pairs of the nitrogen atoms.

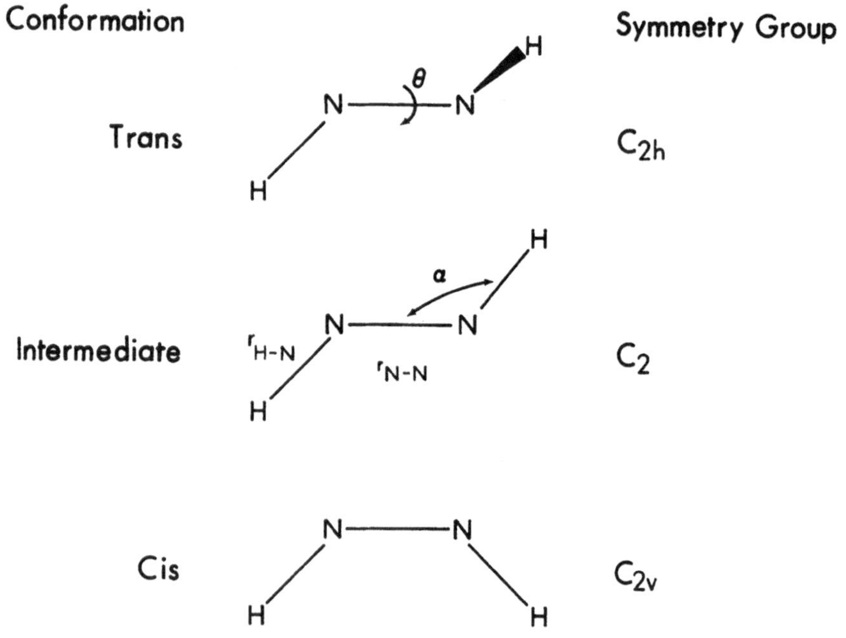

Figure 14.XI. - Geometries of diimine

From this diagram it can be predicted that torsion in the groundstate is a symmetry forbidden process. The configuration $(B)^2 (A)^2 (A)^2$ of the trans isomer is correlated with $(A)^2 (B)^2 (A)^2$ which is an excited cis conformer. The twist will be a symmetry allowed process in the first excited singlet and triplet states. Also the trans configuration $(B)^2 (A)^2 (A)^1 (B)^1$ correlates with the cis $(A)^2 (B)^2 (B)^1 (A)^1$ configuration.

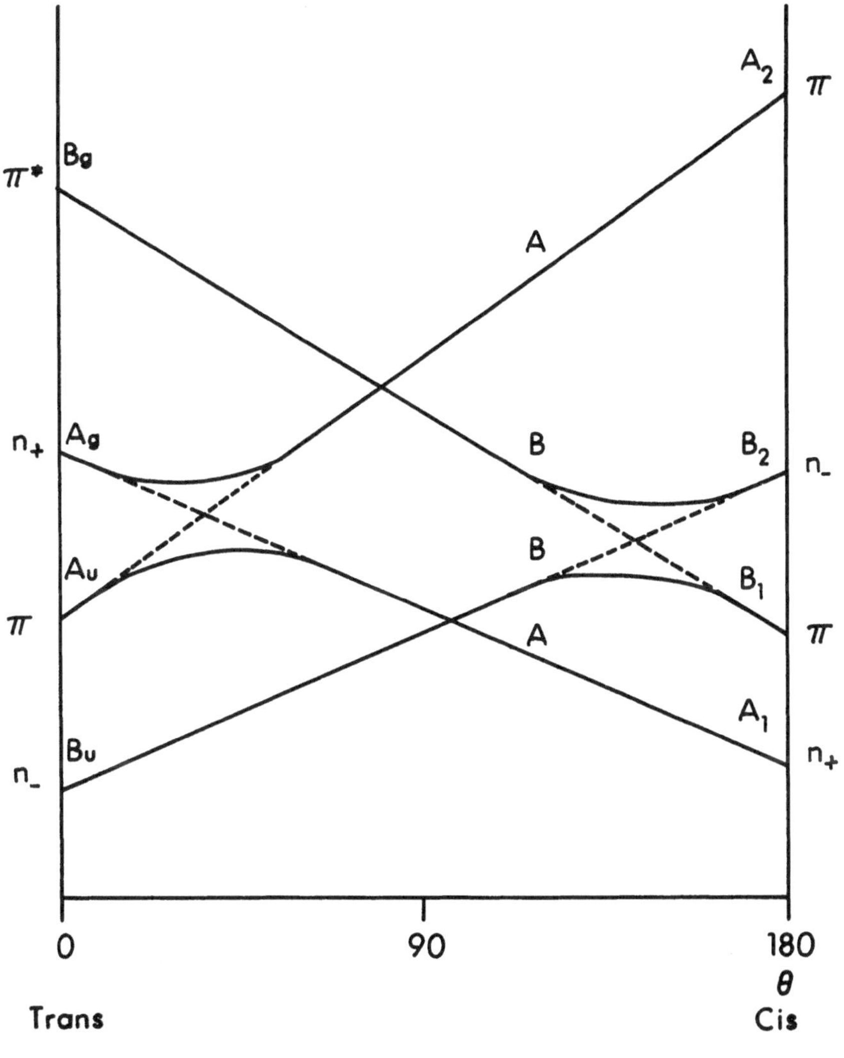

Figure 15.XI. - Correlation diagram for diimine

Figure 16.XI presents the energy surface obtained with the Coupling Operator method using a minimal basis set with diffuse p orbitals. In the intermediate zones the N-N distance increases and the NNH angle decreases. The distance N-N is practically insensitive to torsion for

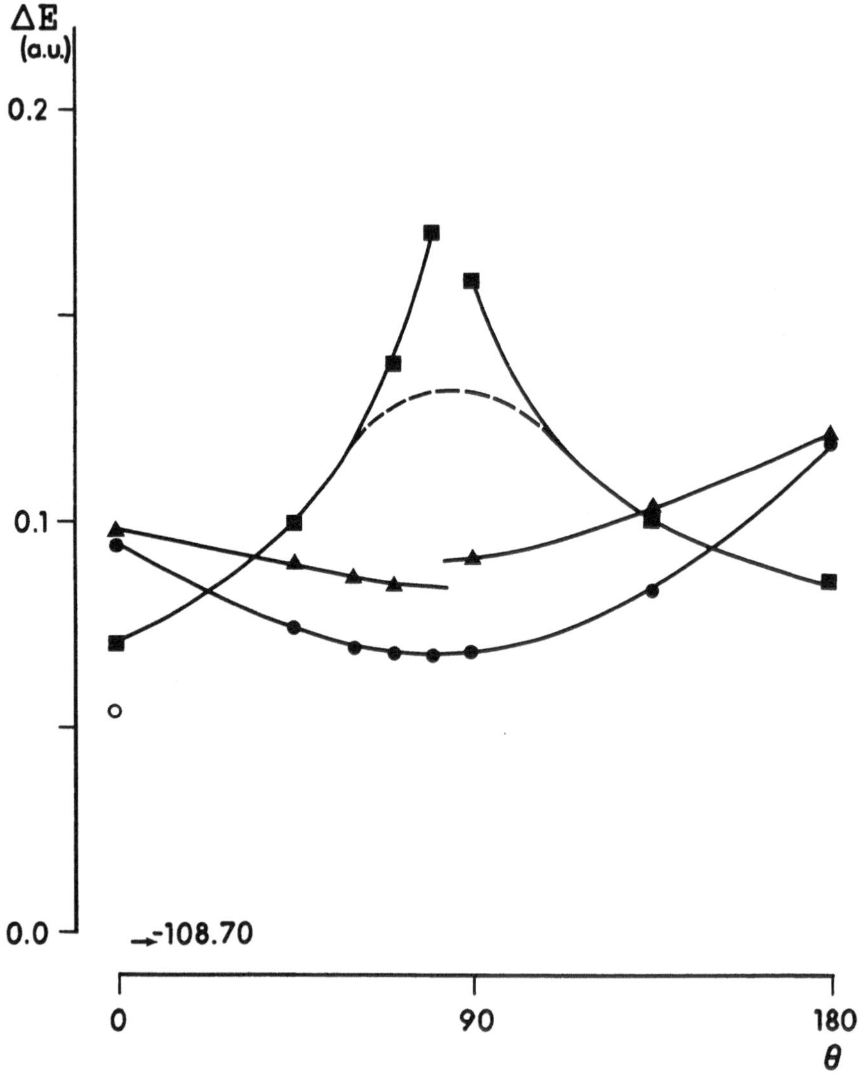

Figure 16.XI. - Molecular potential surface for diimine. Calculation with minimal basis set plus diffuse p orbitals. (■) Groundstate . (●) Triplet state. (▲) Triplet state obtained by means of an EHP method, shown for comparative purposes. (O) Groundstate at optimum geometry

any studied states.
A discontinuity can be observed in the groundstate ; this is precisely due to the symmetry-forbidden processes. The groundstate wavefunction is not the same on both sides of the discontinuity, and the unique manner in which to overcome this difficulty involves the use of a multi-configurational procedure. The discontinuity disappears in the triplet because all the orbitals of A and B type are optimized together.
In order to study this effect, the molecular electostatic potential is constructed for the following cases: 1) Groundstate HOMO and LUMO. 2) Triplet state HOMO and LUMO. In the triplet the optimum geometry has been taken at a rotation angle of 90°. Due to the molecular symmetry, this conformation has the advantage that the calculated plane is equal to another perpendicular one which contains the N-N bond. The potential has been calculated by supposing a positive charge exists on each nitrogen and that two electrons occupy the studied orbital. This situation is totally arbitrary, but is sufficient for comparative purposes. The electrostatic potential maps for groundstate and triplet cases are presented in Figure 17.XI. The triplet exhibits a deep minimum for both orbitals.

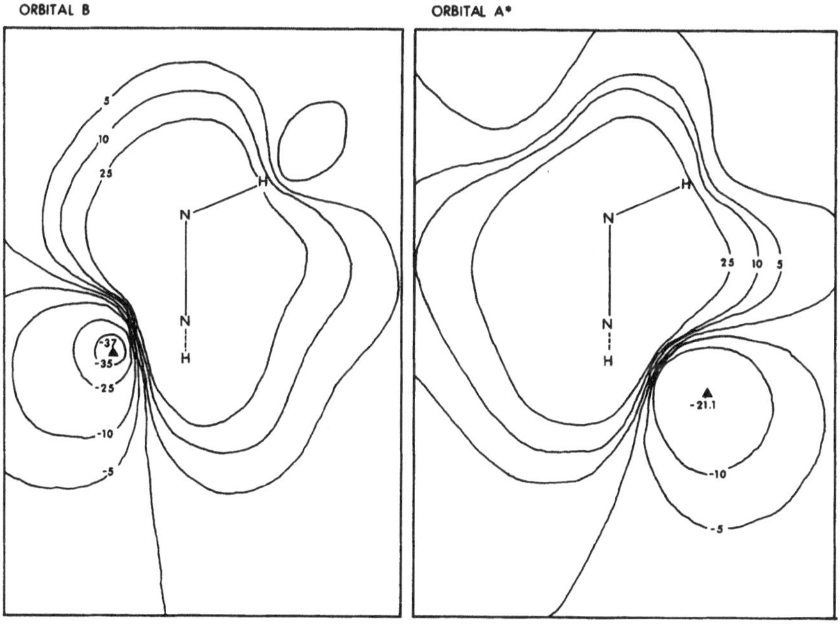

Figure 17.XI.a. - Electrostatic potential for HOMO and LUMO of diimine. Minimal basis set plus diffuse p orbitals. Groundstate

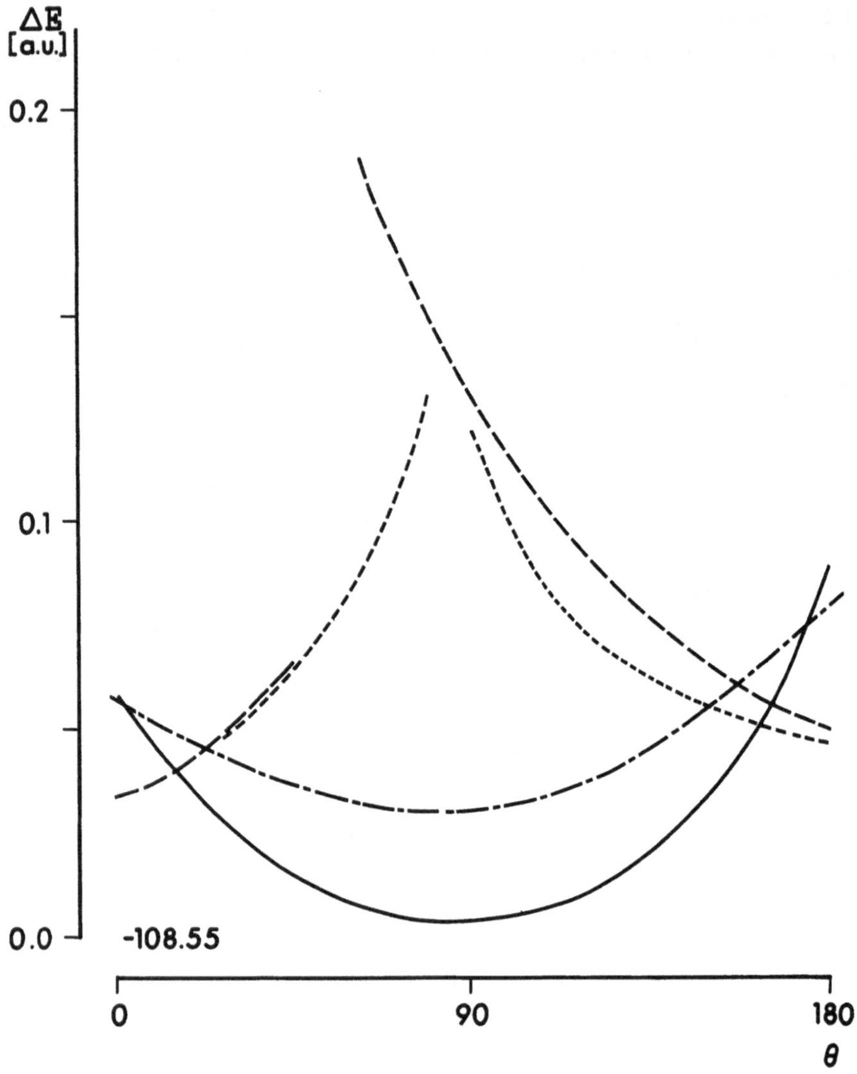

Figure 18.XI. - Potential energy of diimine.

a) (– – –) Groundstate , minimal basis set

b) (- - - -) Groundstate , minimal basis set and diffuse p orbitals.

c) (—·—) Triplet state, minimal basis set

d) (——) Triplet state, minimal basis set and diffuse p orbitals

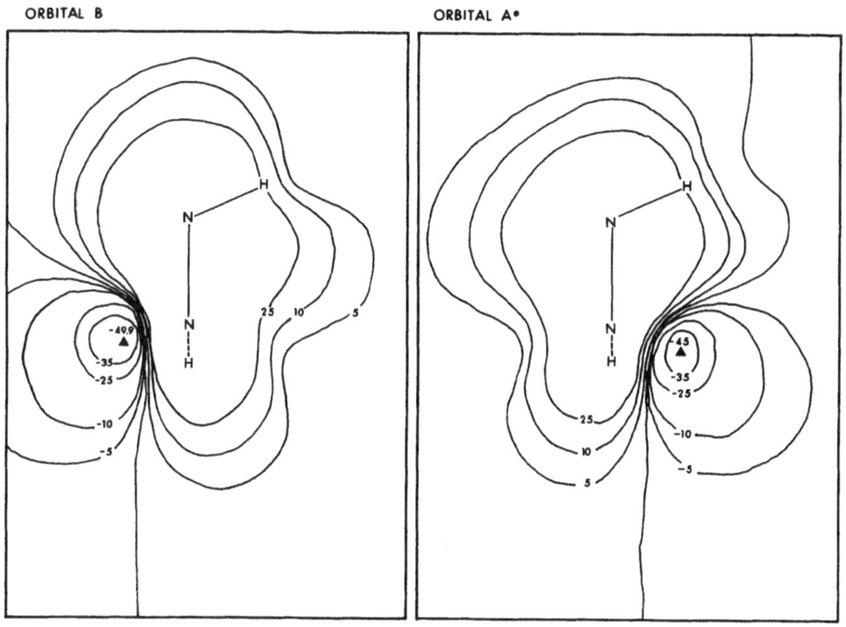

Figure 17.XI.b. - Electrostatic potential for HOMO and LUMO of diimine. Minimal basis set plus diffuse p orbitals. Triplet state

Without considering the energy discontinuity, one should note that the transition energies between triplet and groundstate are too low, this fact being more noticeable in the intermediate situation. This can be explained by the different correlation energy included in the wavefunctions representing each state.

Figure 18.XI presents the results obtained with a minimal basis set. The relative energies found when an augmented basis set is used are also shown. It is assumed that the groundstate in the trans form has the same energy in both basis sets. In this case geometry optimization has not been performed, but the geometries previously obtained have been used. The potential surface is similar in both calculations. The decrease found in the triplet transition energy in the intermediate zone is due to the fact that a poor basis set increases correlation energy differences.

Figure 19.XI presents the electrostatic molecular potential maps of the same orbitals, states and geometry as those studied in the diffuse orbital basis set calculation. The general forms coincide with those

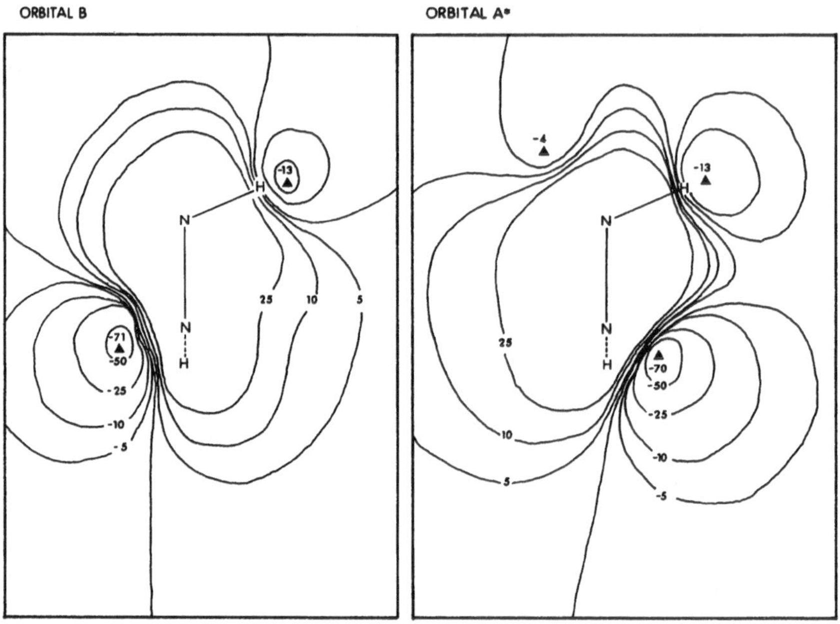

Figure 19.XI.a. - Electrostatic potential of HOMO and LUMO of diimine. Minimal basis set. Groundstate .

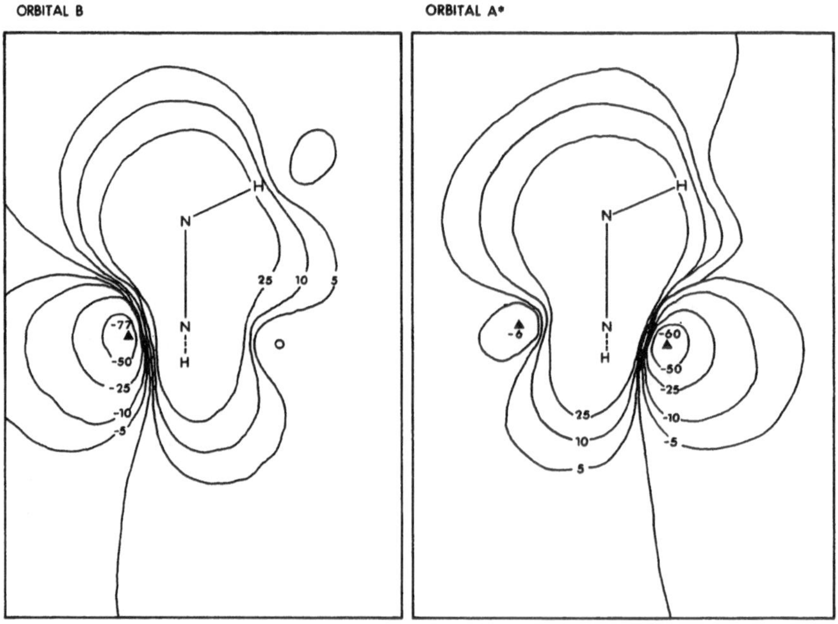

Figure 19.XI.b. - Electrostatic potential of HOMO and LUMO of diimine. Minimal basis set. Triplet state

obtained with the previous basis, however the well depth is more pronounced with the minimal basis. This fact coincides with the general behaviour of the electrostatic molecular potentials.

9. Methylenimine

Due to the small differences observed in the diimine treatment when both basis sets are used, it has been considered sufficient to use a minimal basis set in this case.

The present structure possesses 9 degrees of freedom. Figure 20.XI.a. shows the convention used in order to distinguish them. Actually, only 7 internal coordinates are drawn in the figure. The reason for this omission will be given below. First of all, the groundstate geometry is optimized by the procedure given in the following steps: a) The C-N, N-H and C-H bond lengths are systematically varied as are the angles α and γ. In this step the following constraints are used: the methylene hydrogens are symmetrically maintained with respect to C-N bond at the same distance from the carbon atom. The angle δ is kept constant at 90°, so that the molecule is contained in a unique plane. In order to obtain the optimal geometry from the standard distances and angles two optimizations are needed for each parameter, or in other words, four points for each optimization. It has been considered that the optimal geometry is obtained when any variation of geometrical parameters gives a total energy variation less than 3×10^{-5} a.u. b) The constraints are independently relaxed, that is the energy variation is observed when only one constraint is relaxed. In all cases a rise in

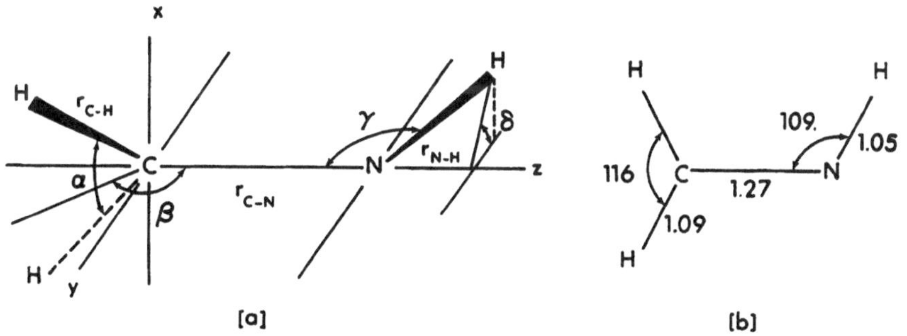

Figure 20.XI. - Molecular geometries of methylenimine (a) Degrees of freedom. (b) Optimum Geometry of groundstate

energy has been observed, except when the distances C-H are varied and, as a consequence, an asymmetric methylene group has been obtained. The difference between distances is less than 0.001 Å and the energy varies less than 1.10^{-6} a.u.; since there is so little difference, this kind of symmetry has not been taken into account. The optimal geometry is shown in Figure 20.XI.b.

On the other hand, the qualitative study of the triplet state has been carried out in a manner similar to the diimine case, but with the difference that throughout C-N bond rotation no symmetry has been conserved. Although Woodward-Hoffmann's rules cannot be used in this case, the results are expected to be very close to the diimine case.

The procedure followed in obtaining the potential surface is slightly different from that in the diimine molecule. The starting point is the triplet state with a ground state geometry, that is, a vertical transition is performed. Experimentally it is difficult to carry on this process for the triplet, but not for the singlet, however we use the fact that the behaviour of both excited states should be practically the same. The computational procedure cannot be used in the excited singlet because the symmetry of this state coincides with the ground-state symmetry and the variational principle does not hold.

At present, the geometrical parameters are optimized using a gradient technique. The results are presented in two parts: 1) Figure 21.XI shows the energy variation vs. the stretching of the C-N bond length, which is approximately the energy gradient direction, except in the minimum neighbourhood. In the figure are shown some points numbered from 1 to 8; each point has been obtained by introducing as initial vectors of the SCF process those of the previous number except point 1, which has been obtained with the groundstate vectors. This computational scheme allows for the search of a triplet crossing. The $^3(n-\pi^*)$ state presents a lesser stretching on the C-N bond than the $^3(\pi-\pi^*)$ state, as is expected. A dotted line in fig.21.XI indicates the position of the ground triplet when symmetry has been broken by means of a methylene piramidalization at 179° (it is not possible to compute the higher triplet, because it has the same symmetry as the ground triplet). Figure 22.XI presents the energy variation vs. the torsion angle, and the dotted line represents the boundary of the groundstate evolution, following the gradient direction for the triplet. The full line represents the lowest triplet state energy for a C-N bond length of 1.5244 Å and on an angle β of 150°. The upper line corresponds to the groundstate . The three points marked with (Δ) correspond to the triplet energy when the molecular geometry has been totally optimized.

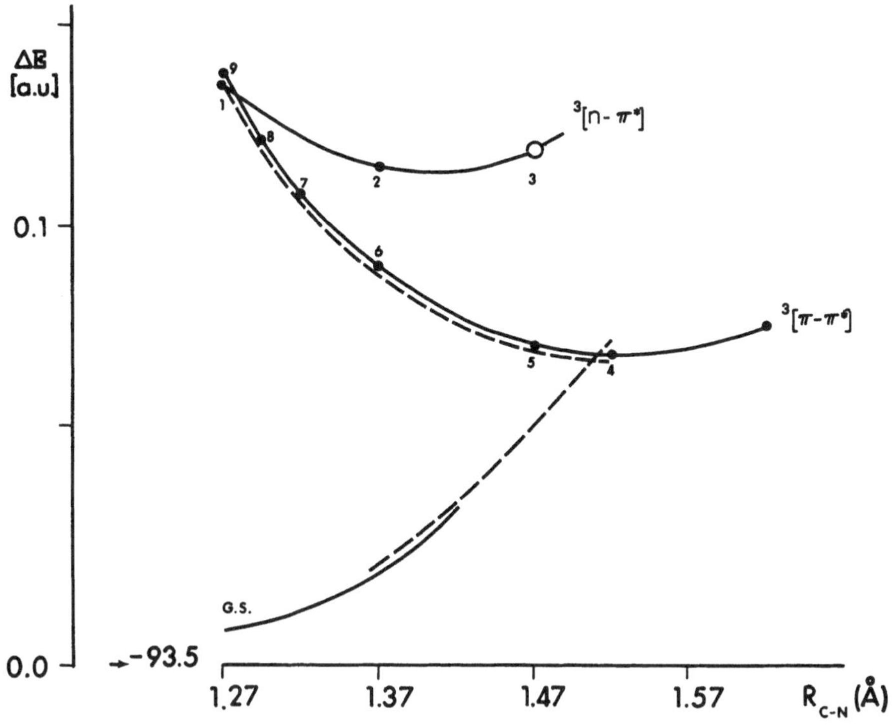

Figure 21.XI. - Energy of methylenimine vs. stretching of C-N bond.
See text for more details

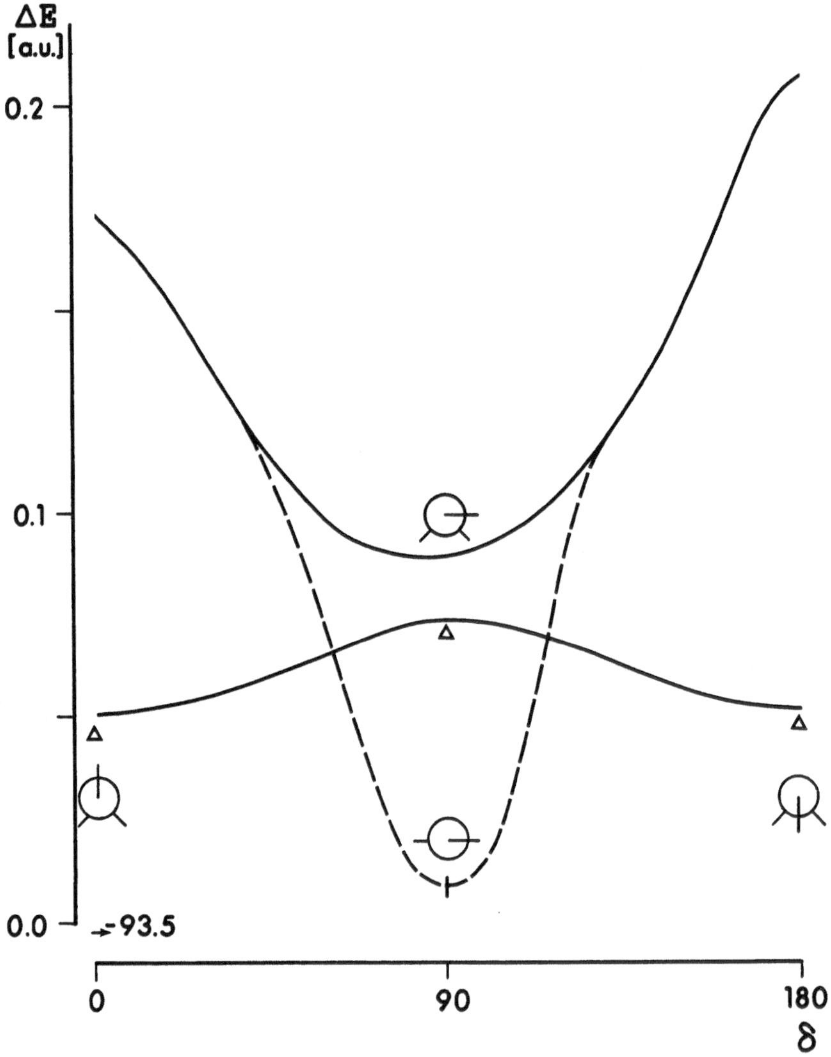

Figure 22.XI. - Potential energy of methylenimine upon rotation on C-N bond. Minimal basis set (see text for more details)

Table 11.XI

Diimine and methylenimine comparative results

a) Diimine

State	R(N-N) (Å)	R(N-H) (Å)	α(NNH) (°)	ΔE(a.u)[+] mbs+p	mbs
Ground	1.21	1.01	109	-	-
Triplet					
trans	1.37	1.05	118	0.04	0.02
cis	1.30	1.04	133	0.07	0.06
80°	1.41	1.07	111	0.01	-0.03

b) Methylenimine*

State	R(C-N) (Å)	α (°)	β (°)	γ (°)	ΔE[+] (a.u.)
Ground	1.27	116	0	109	-
Triplet					
planar	1.53	115	142	103	0.063
skew	1.47	117	149	105	0.040
staggered	1.47	116	142	105	0.038

c) Comparative bondlengths and rotation barriers

Molecule	R(A-B)(Å) Ground	Triplet	Rotation barrier (kcal/mol) Ground	Triplet
Ethylene	1.34	1.50	139	16
Formaldehyde	1.22	1.46	-	-
Diimine	1.21	1.41	86	34
Methylenimine	1.27	1.47	103	62

(*) R(C-H) = 1.09 Å; R(N-H) = 1.05 Å in all cases.

(+) Vertical transitions.

Discussion on Diimine and Methylenimine

Generally speaking, monoconfigurational methods do not give good results for all the conformations involved in a potential surface calculation. Taking this limitation into account, the SCF method does not present discontinuities other than those due to the chosen wavefunction. The biggest inconvenience in the evaluation of the transition energies results from the different correlation energies between states of different multiplicities.

The general conclusions which can be obtained from the calculations are: a) In the triplet state the distance between the double bond atoms has become augmented considerably, up to single bond values. b) The NNH angle of the diimine, and the piramidalization angle for the methylenimine, play a similar role in the triplet: its variation causes the molecule to adopt a biradical structure. c) The other geometrical parameters are insensitive to the electronic structure variation. It should also be noted that the results with minimal basis set are similar to those obtained with an augmented basis with diffuse p orbitals. In the methylenimine case the near degeneracy between $^3(n-\pi^*)$ and $^3(\pi-\pi^*)$ in the vertical transitions can invoke a crossing which tends towards a geometry very near to the one of groundstate .

The results obtained are comparable to those for other double bond molecules. Ethylene and formaldehyde are chosen for this purpose, because they are the simplest compounds which possess a double bond. First of all, in the triplet states a trend towards bond stretching is observed, which practically equals a single bond in length. The rotational barriers decrease from ethylene to diimine for the groundstate . Although the results are far from the experimental ones, they can be compared because in all cases a minimal basis set has been used. The triplet rotational barrier is interpreted as the energy barrier which the molecule, in the triplet state and at the optimal geometry, needs to exceed for rotation about the double bond by 180°. The results show a tendency for isomerization through a triplet state which is emphasized by the C-N bond. Table 11.XI resumes these results.

10. Glycine

The ground singlet and triplet states of glycine in Zwitterion form have been studied with a double zeta quality basis set. The geometry has been obtained from crystal structure data and is shown in Table 12.XI. The molecular parameters for both states can be found in

Table 13.XI. The electrostatic potential map of the molecule has been

Table 12.XI

Glycine Zwitterion: coordinates (a.u.)(*)

	X	Y	Z
O_1	2.0953	0.0	3.9745218
O_2	-2.1045659	0.0	3.9607282
C_1	0.0	0.0	2.8837777
C_2	0.0	0.0	0.0
N	2.588004	0.0	-1.0403709
H_1	2.5949567	0.0	-3.0321686
H_2	3.6251923	-1.6640061	-1.0435612
H_3	3.5480763	1.68614	-1.0592724
H_4	-0.4375792	1.7985716	-0.8785641
H_5	-0.4323904	-1.7636194	-0.9536459

(*) Molecular bond distances and angles from: P.G. Jönson, A. Kvick; Acta Crystallographica, B28, 1827 (1972).

calculated by means of the approximation described in section 1.4.VIII. Figures 23.XI, and 24.XI display the electrostatic potential surface for singlet and triplet, respectively, computed in a plane situated 2.5 a.u. above the plane which contains the heavy atoms of the molecule (XZ). The differences between the two states are notable. Groundstate singlet has two well defined minima corresponding to the oxygen lone pairs. Triplet state, on the contrary, has practically lost these minima. This behaviour is consistent with the charge transfer from the whole molecule to the hydrogens of the NH_3^+ group.

Table 13.XI

Glycine Zwitterion molecular parameters (a.u.)

State	Energy	μ	$\langle r^2 \rangle$
^1A	-282.61302	5.700	420.70
^3A	-282.44235	0.410	438.13

	$\langle r^{-1} \rangle$		$\langle \delta \rangle$		Q	
State	^1A	^3A	^1A	^3A	^1A	^3A
O_1	-22.445	-22.248	295.37	295.83	-0.5882	-0.1975
O_2	-22.453	-22.239	295.41	295.83	-0.5408	-0.1826
C_1	-14.632	-14.523	119.71	119.68	0.3681	0.3094
C_2	-14.632	-14.643	119.78	119.66	-0.0434	0.0031
N	-18.174	-18.303	194.27	194.92	-0.5344	-0.2993
H_1	-0.840	-0.949	0.3520	0.309	0.3011	-0.1591
H_2	-0.840	-0.938	0.3312	0.320	0.3655	0.1084
H_3	-0.840	-0.940	0.3386	0.328	0.3574	0.1072
H_4	-1.002	-1.013	0.3813	0.366	0.1560	0.1533
H_5	-1.001	-1.012	0.3808	0.364	0.1585	0.1570

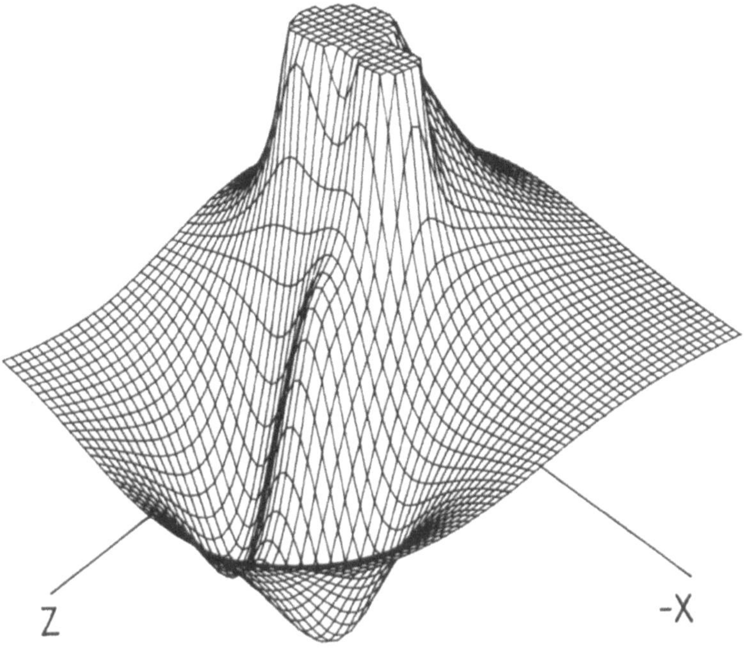

Figure 23.XI. - Glycine electrostatic molecular potential at 2.5 a.u. from XZ plane. Groundstate

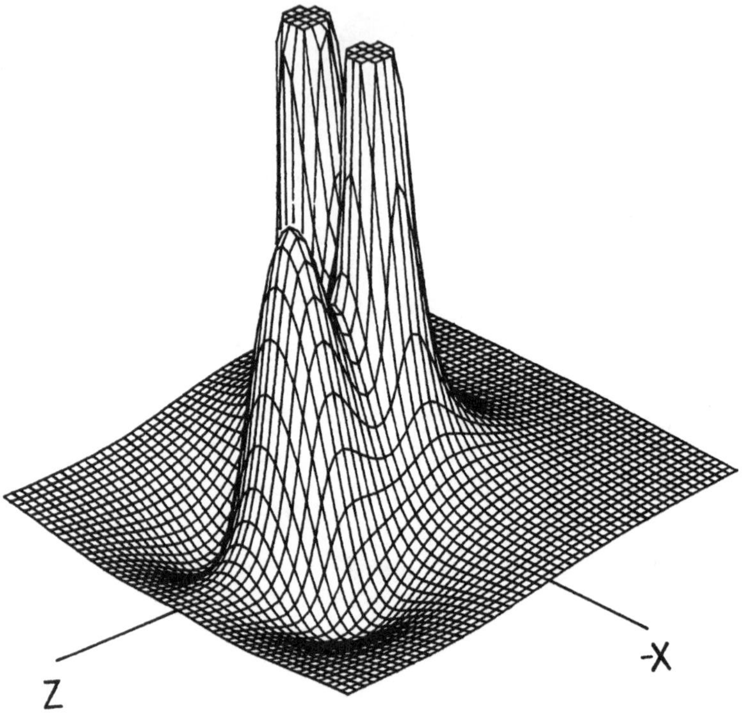

Figure 24.XI. - Glycine electrostatic molecular potential at 2.5 a.u. from XZ plane. Triplet state

11. Excited States of Some Molecules with C=O and C=N Bonds: INDO Procedures

The geometry of the excited states for formaldehyde, formic acid, formamide, methylenimine, ketene, and diazomethane have been studied in a monoconfigurational formalism in order to test the generalized SCF procedure in an empirical framework. The INDO approach has been chosen for this purpose (see section 3.VIII).

Table 14.XI shows the values of the out of plane bending angles obtained for the excited states of the molecules considered here. For formaldehyde, formic acid, formamide and methylenimine the angle studied has been the one associated with the C=X bond. In ketene and diazomethane, the terminal C=O and N=N bonds have been taken as parameters.

Experimental values for formaldehyde triplet give an angle of 35° and for the singlet experimental values range from 21° to 31°. For the singlet in formic acid the experimental value is 32°. Although experimental information is scarce, one can conclude that the computed values can be considered quantitatively in the range of expected values, although slightly overestimated. The transition energies for the studied molecules are given in Table 15.XI, giving reasonable values, when compared with experimental and other theoretical sources.

As a consequence, one can say that the outlined Coupling Operator procedures can be used with empirical methods in order to obtain good qualitative descriptions of the excited states.

Table 14.XI

Out of plane bending angles for excited singlets and triplets of some molecules: INDO approximation [a]

Molecule	State	Angle (°)
Formaldehyde	$^3A''$ $(n-\pi^*)$	37
	$^1A''$ $(n-\pi^*)$	34
Formic acid	$^3A''$ $(n-\pi^*)$	39
	$^1A''$ $(n-\pi^*)$	33
Formamide	^3A $(n-\pi^*)$	37
Methylenimine	$^3A''$ $(n-\pi^*)$	20
	$^1A''$ $(n-\pi^*)$	20
Ketene	$^3A''$ $(\pi-\pi^*)$	47
	$^1A''$ $(\pi-\pi^*)$	43
Diazomethane	$^3A''$ $(\pi-\pi^*)$	28
	$^1A''$ $(\pi-\pi^*)$	22

[a] From: R. Caballol; Tesis Doctoral, Instituto Quimico de Sarria (1976).

Table 15.XI

Vertical transition energies for some molecules. INDO approximation

Molecule	State	Energy (eV) calculated*	Experimental (eV)
Formaldehyde	$^3A_2$ (n-π*)	3.40	3.13 - 3.44[a]
	$^1A_2$ (n-π*)	3.74	3.51 - 5.39[a]
Formic acid	$^3A''$ (n-π*)	3.67	-
	$^1A''$ (n-π*)	4.21	4.8 - 5.7[a]
Formamide	$^3A''$ (n-π*)	4.09	-
	$^1A''$ (n-π*)	4.30	5.65[b]
Methylenimine	$^3A''$ (n-π*)	4.39	-
	$^1A''$ (n-π*)	4.96	-
Ketene	$^3A_2$ (π-π*)	4.60	-
	$^1A_2$ (π-π*)	4.40	3.34 - 4.77[a]
Diazomethane	$^3A_2$ (π-π*)	2.35	2.61 - 3.87[a]
	$^1A_2$ (π-π*)	2.50	2.48 - 4.18[c]

[a] G. Herzberg; "Molecular Spectra and Molecular Structure", Vol. III, Van Nostrand, Princeton (1967).

[b] T.L. Whitten, M. Hackmeyer; J. Chem. Phys., 56, 5584 (1969).

[c] F.W. Kirbride, R.G.W. Norrish; J. Chem. Soc., 119 (1933).

* From: R. Caballol, Tesis Doctoral, Instituto Quimico de Sarria (1976).

Appendix A

Monoconfigurational State Parameters

In the following tables are given the state parameters $\{\omega_i\}$; $\{\alpha_{ij}\}$ and $\{\beta_{ij}\}$, associated with an energy expression like:

$$E = \sum_i \omega_i h_{ii} + \sum_i \sum_j (\alpha_{ij} J_{ij} - \beta_{ij} K_{ij}),$$

for some selected electronic configurations. Only the state parameters for each shell are presented here. A shell is defined as a MO subset which shares the same Fock Operator. (See section 2.IX).
When more than one wavefunction can be attached to a given state, an average of the corresponding energies has been used.
The following examples will illustrate the construction of the tables.

1. Triplet and Singlet: Nondegenerate MO

The energy for these states can be written as

$$\begin{aligned} E = & \sum_{i\varepsilon C} 2\, h_{ii} + h_{kk} + h_{\ell\ell} \\ & + \sum_{i\varepsilon C} \sum_{j\varepsilon C} (2\, J_{ij} - K_{ij}) + \sum_{i\varepsilon C} [(2\, J_{ik} - K_{ik}) + (2\, J_{i\ell} - K_{i\ell})] \\ & + J_{k\ell} \pm K_{k\ell} \end{aligned}$$

where the minus sign stands for the triplet and the plus sign for the singlet.
In the triplet case, $J_{k\ell} - K_{k\ell}$ can be written as

$$J_{k\ell} - K_{k\ell} = \frac{1}{2} \sum_{p\varepsilon O} \sum_{q\varepsilon O} (J_{pq} - K_{pq})$$

where $O = \{k, \ell\}$, and also

$$\begin{aligned} E = & \sum_{i\varepsilon C} 2\, h_{ii} + \sum_{p\varepsilon O} h_{pp} + \sum_{i\varepsilon C} \sum_{j\varepsilon C} (2\, J_{ij} - K_{ij}) \\ & + \frac{1}{2} \sum_{i\varepsilon C} \sum_{p\varepsilon O} (2\, J_{ip} - K_{ip}) + \frac{1}{2} \sum_{i\varepsilon C} \sum_{p\varepsilon O} (2\, J_{pi} - K_{pi}) \\ & + \frac{1}{2} \sum_{p\varepsilon O} \sum_{q\varepsilon O} (J_{pq} - K_{pq}). \end{aligned}$$

So there are only two shells needed to construct Fock operators and energy: the closed shell C, containing all the doubly occupied MO's

and the open shell O, which contains the singly occupied MO's $\{k,\ell\}$, Table A.3.2 gives this situation.
Three shells shall be defined in the singlet case: A closed shell C, which has the same MO's as in the triplet case, and two shells for each singly occupied MO: k and ℓ. Table A 3.1 gives the corresponding state parameters. From the table it is easy to construct the Fock operators for this situation:
Taking

$$F = h + \sum_{j \varepsilon C} (2\, J_j - K_j),$$

then

a) $(\forall\ i\ \varepsilon\ C)$: $$F_C = F + \frac{1}{2}\, [(2\, J_k - K_k) + (2\, J_\ell - K_\ell)]$$

b) $$F_k = \frac{1}{2}\, F + \frac{1}{2}(J_\ell + K_\ell)$$

c) $$F_\ell = \frac{1}{2}\, F + \frac{1}{2}(J_k + K_k).$$

2. Doublet State in a Twofold Degenerate Orbital

There are two possible orbital occupancies

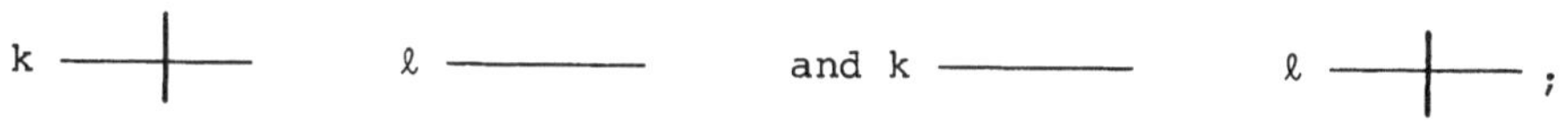

the energies for each situation are

$$E_k = \sum_{i \varepsilon C} 2\, h_{ii} + h_{kk} + \sum_{i \varepsilon C} \sum_{j \varepsilon C} (2\, J_{ij} - K_{ij}) + \sum_{i \varepsilon C} (2\, J_{ik} - K_{ik})$$

and

$$E_\ell = \sum_{i \varepsilon C} 2\, h_{ii} + h_{\ell\ell} + \sum_{i \varepsilon C} \sum_{j \varepsilon C} (2\, J_{ij} - K_{ij}) + \sum_{i \varepsilon C} (2\, J_{i\ell} - K_{i\ell}).$$

In order to preserve the orbital symmetry an average expression is used

$$E = \frac{1}{2}(E_k + E_\ell) = \sum_{i \varepsilon C} 2\, h_{ii} + \frac{1}{2}\, h_{kk} + \frac{1}{2}\, h_{\ell\ell}$$

$$+ \sum_{i \varepsilon C} \sum_{j \varepsilon C} (2\, J_{ij} - K_{ij}) + \frac{1}{2} \sum_{i \varepsilon C} [2\, J_{ik} - K_{ik}) + (2\, J_{i} - K_{i\ell})]$$

which conveniently arranged gives Table B 1. This is a case where, like the triplet discussed before, only the two shells are needed; the Closed Shell C containing the doubly occupied MO's and the Open Shell O, which contains both degenerate orbitals. The Fock Operators are in this case

a) $(\forall\ i \in C)$: $F_C = F + \frac{1}{2} \sum_{p \in O} (2 J_p - K_p)$

b) $(\forall\ p \in O)$: $F_O = \frac{1}{2} F$,

which conserve F, the same expression as used in the singlet case.

3. A Degenerate Case

Let's suppose that we are facing an electronic configuration like:

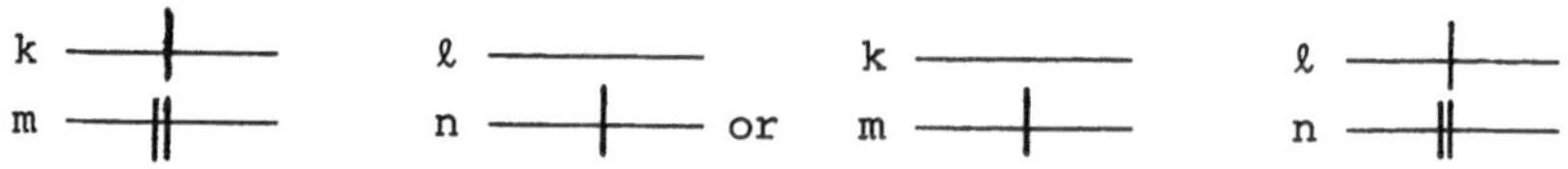

as illustrated in Table B 7.1. This case will give in a linear molecule two states Σ^+ and Σ^-. In order to simplify the calculation only these four orbitals will be considered. Four possible determinants can be constructed as a basis set for the singlet state wavefunctions, namely

$$\Phi_1 = |(m\alpha)(m\beta)(n\alpha)(k\beta)|$$
$$\Phi_2 = |(m\alpha)(m\beta)(n\beta)(k\alpha)|$$
$$\Phi_3 = |(n\alpha)(n\beta)(m\alpha)(\ell\beta)|$$
$$\Phi_4 = |(n\alpha)(n\beta)(m\beta)(\ell\alpha)|.$$

from here two wavefunctions can be constructed

$$\Psi_\pm = \frac{1}{2}[(\Phi_1 - \Phi_2) \pm (\Phi_3 - \Phi_4)],$$

so the energies can be calculated as

$$E_\pm = \langle\Psi_\pm|\mathcal{H}|\Psi_\pm\rangle = \frac{1}{4} \sum_{i,j} s_i s_j \langle\Phi_i|\mathcal{H}|\Phi_j\rangle,$$

where the symbols $\{s_i\}$ represent the proper sign of function Φ_i in the $\Psi_\pm$ combination. In order to obtain the expression for $E_\pm$, the following matrix elements should be known (see Appendix B):

a) $$\langle\Phi_1|\mathcal{H}|\Phi_1\rangle = 2h_{mm} + h_{nn} + h_{kk} + 2J_{mm} - K_{mm} + 2J_{mn} - K_{mn} + 2J_{mk} - K_{mk} + J_{nk}$$

b) $\langle\Phi_2|\mathcal{H}|\Phi_2\rangle = \langle\Phi_1|\mathcal{H}|\Phi_1\rangle$

c) $\langle\Phi_3|\mathcal{H}|\Phi_3\rangle = 2h_{nn} + h_{mm} + h_{\ell\ell} + 2J_{nn} - K_{nn} + 2J_{mn} - K_{mn} + 2J_{n\ell}$

$- K_{n\ell} + J_{m\ell}$

d) $\langle\Phi_4|\mathcal{H}|\Phi_4\rangle = \langle\Phi_3|\mathcal{H}|\Phi_3\rangle$

e) $\langle\Phi_1|\mathcal{H}|\Phi_2\rangle = - K_{nk}$

f) $\langle\Phi_3|\mathcal{H}|\Phi_4\rangle = - K_{m\ell}$

g) $\langle\Phi_1|\mathcal{H}|\Phi_3\rangle = (m\ell|nk) - (mn|k\ell)$

h) $\langle\Phi_1|\mathcal{H}|\Phi_4\rangle = - (m\ell|nk)$

i) $\langle\Phi_2|\mathcal{H}|\Phi_4\rangle = (m\ell|nk)$

j) $\langle\Phi_2|\mathcal{H}|\Phi_3\rangle = (m\ell|nk) - (mn|k\ell)$.

We have for $E_\pm$

$$\begin{aligned}
E_\pm = {} & \tfrac{3}{2}(h_{mm} + h_{nn}) + \tfrac{1}{2}(h_{kk} + h_{\ell\ell}) \\
& + \tfrac{1}{2}[(2J_{mm} - K_{mm}) + (2J_{nn} - K_{nn})] + 2J_{mn} - K_{mn} \\
& + \tfrac{1}{2}[(2J_{mk} - K_{mk}) + (2J_{n\ell} - K_{n\ell})] \\
& + \tfrac{1}{2}[(J_{m\ell} + K_{m\ell}) + (J_{nk} + K_{nk})] \\
& \pm [2(m\ell|nk) - (mn|k\ell)].
\end{aligned}$$

The last term takes into account the interaction between $(\Phi_1 - \Phi_2)$ and $(\Phi_3 - \Phi_4)$, and does not appear in the averaged coefficients of Table B 7.1. In this sense, neglecting this term, we have $E_+ = E_-$.
So, care should be taken when using the parameters given in the tables for the degenerate cases. Interactions between degenerate functions have not been taken into account, in order to preserve the monoconfigurational structure of the energy expression.
If the interaction terms should be included in the SCF procedure, a set of crossed term Fock operators linking m, n, k and ℓ will be present

and in this case a more complicated Coupling Operator will appear. The triplet functions have similar properties. In the energy would appear a term like

$$\frac{1}{2}(J_{m\ell} - K_{m\ell} + J_{nk} - K_{nk})$$

and the interaction would be $\pm(mn|k\ell)$.

Tables of State Parameters

A. Nondegenerate open shell orbitals

A1. Closed shell: Singlet state

	ω	α	C	β	C
C	2		2		1

A2. One open shell singly occupied: Doublet state

	ω	α	C	k	β	C	k
C	2		2	1		1	1/2
k	1		1	0		1/2	0

A3. Two singly occupied orbitals

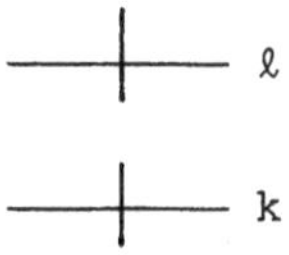

A3.1 Two open shells: Singlet state

	ω	α	C	k	ℓ	β	C	k	ℓ
C	2		2	1	1		1	1/2	1/2
k	1		1	0	1/2		1/2	0	-1/2
ℓ	1		1	1/2	0		1/2	-1/2	0

A3.2 One open shell: Triplet state

$k, \ell \in O$

	ω	α	C	O	β	C	O
C	2		2	1		1	1/2
O	1		1	1/2		1/2	1/2

A4. Three singly occupied orbitals

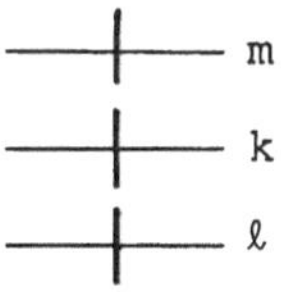

A4.1 One open shell: Quadruplet state

with m, k, ℓ ε O the same as in A3.2.

A4.2 Three open shells: Two doublet states

A4.2.1 (Doublet 1)

	ω	α	C	k	ℓ	m	β	C	k	ℓ	m
C	2		2	1	1	1		1	1/2	1/2	1/2
k	1		1	0	1/2	1/2		1/2	0	1/2	-1/4
ℓ	1		1	1/2	0	1/2		1/2	1/2	0	-1/4
m	1		1	1/2	1/2	0		1/2	-1/4	-1/4	0

A4.2.2 (Doublet 2)

	ω	α	C	k	ℓ	m	β	C	k	ℓ	m
C	2		2	1	1	1		1	1/2	1/2	1/2
k	1		1	0	1/2	1/2		1/2	0	1/4	1/4
ℓ	1		1	1/2	0	1/2		1/2	1/4	0	-1/2
m	1		1	1/2	1/2	0		1/2	1/4	-1/2	0

B. Twofold degenerate open shell orbitals

B1. Two degenerate orbitals and one electron

k  ℓ k, ℓ ε O Doublet state

	ω	α	C	O	β	C	O
C	2		2	1/2		1	1/4
O	1/2		1/2	0		1/4	0

B2. Two degenerate orbitals and two electrons

The same parameters as in A3.1 and A3.2 with

k  ℓ 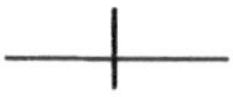

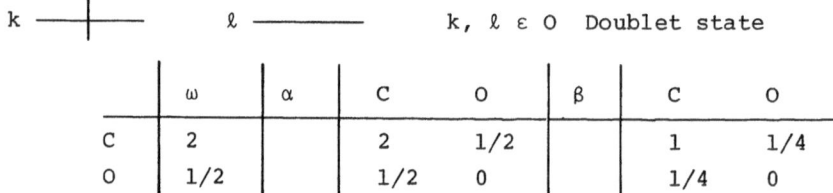

B3. Two degenerate orbitals and three electrons

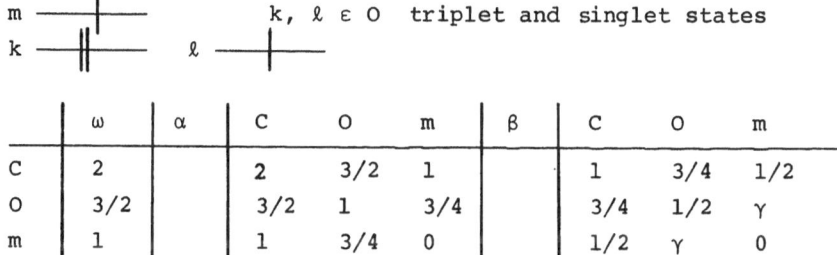

k, ℓ ε O Doublet state

	ω	α	C	O	β	C	O
C	2		2	3/2		1	3/4
O	3/2		3/2	1		3/4	1/2

B4. Three orbitals, two degenerate and two electrons

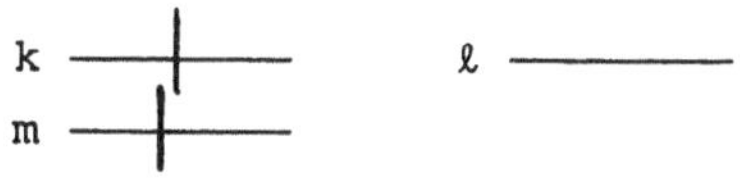

k, ℓ ε O Triplet and singlet states

	ω	α	C	m	O	β	C	m	O
C	2		2	1	1/2		1	1/2	1/4
m	1		1	0	1/4		1/2	0	γ
O	1/2		1/2	1/4	0		1/4	γ	0

For the triplet γ = 1/4

For the singlet γ = - 1/4

B5. Three orbitals, two degenerate and four electrons

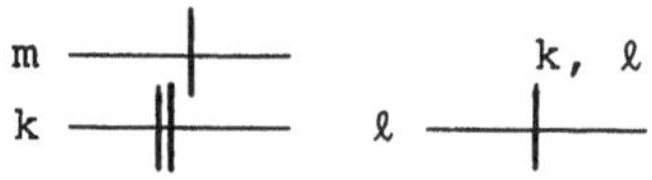

k, ℓ ε O triplet and singlet states

	ω	α	C	O	m	β	C	O	m
C	2		2	3/2	1		1	3/4	1/2
O	3/2		3/2	1	3/4		3/4	1/2	γ
m	1		1	3/4	0		1/2	γ	0

For the triplet γ = 1/2

For the singlet γ = 0

B6. Four orbitals, degenerate two by two and two electrons (one in each orbital)

B6.1

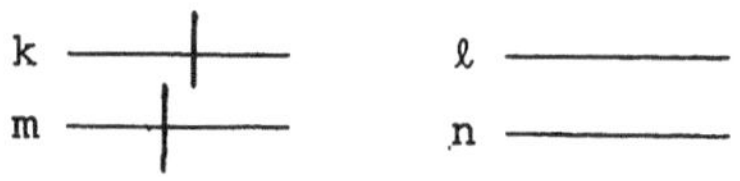

Triplet and singlet states

	ω	α	C	m	n	k	ℓ	β	C	m	n	k	ℓ
C	2		2	1/2	1/2	1/2	1/2		1	1/4	1/4	1/4	1/4
m	1/2		1/2	0	0	1/4	0		1/4	0	0	γ	0
n	1/2		1/2	0	0	0	1/4		1/4	0	0	0	γ
k	1/2		1/2	1/4	0	0	0		1/4	γ	0	0	0
ℓ	1/2		1/2	0	1/4	0	0		1/4	0	γ	0	0

For the triplet $\gamma = 1/4$
For the singlet $\gamma = -1/4$

B6.2

Triplet and singlet states

	ω	α	C	m	n	k	ℓ	β	C	m	n	k	ℓ
C	2		2	1/2	1/2	1/2	1/2		1	1/4	1/4	1/4	1/4
m	1/2		1/2	0	0	0	1/4		1/4	0	0	0	γ
n	1/2		1/2	0	0	1/4	0		1/4	0	0	γ	0
k	1/2		1/2	0	1/4	0	0		1/4	0	γ	0	0
ℓ	1/2		1/2	1/4	0	0	0		1/4	γ	0	0	0

For the triplet $\gamma = 1/4$
For the singlet $\gamma = -1/4$

B7. Four orbitals, degenerate two by two and four electrons (three on the first degenerate pair and one in the second)

B7.1

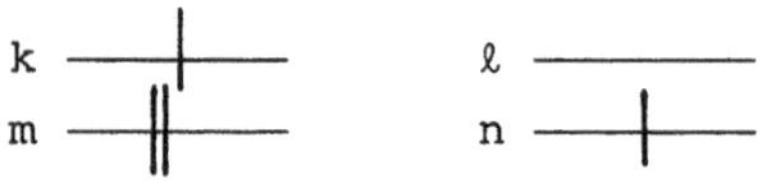

Triplet and singlet states

	ω	α	C	m	n	k	ℓ	β	C	m	n	k	ℓ
C	2		2	3/2	3/2	1/2	1/2		1	3/4	3/4	1/4	1/4
m	3/2		3/2	1	1	1/2	1/4		3/4	1/2	1/2	1/4	γ
n	3/2		3/2	1	1	1/4	1/2		3/4	1/2	1/2	γ	1/4
k	1/2		1/2	1/2	1/4	0	0		1/4	1/4	γ	0	0
ℓ	1/2		1/2	1/4	1/2	0	0		1/4	γ	1/4	0	0

For the triplet $\gamma = 1/4$
For the singlet $\gamma = -1/4$

B7.2

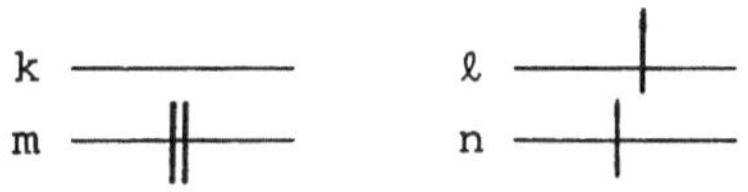

Triplet and singlet states

	ω	α	C	m	n	k	ℓ	β	C	m	n	k	ℓ
C	2		2	3/2	3/2	1/2	1/2		1	3/4	3/4	1/4	1/4
m	3/2		3/2	1	1	1/4	1/2		3/4	1/2	1/2	γ	1/4
n	3/2		3/2	1	1	1/2	1/4		3/4	1/2	1/2	1/4	γ
k	1/2		1/2	1/4	1/2	0	0		1/4	γ	1/4	0	0
ℓ	1/2		1/2	1/2	1/4	0	0		1/4	1/4	γ	0	0

For the triplet $\gamma = 1/4$

For the singlet $\gamma = -1/4$

B8. Four orbitals, degenerate two by two, and six electrons, (three in each degenerate pair)

B8.1

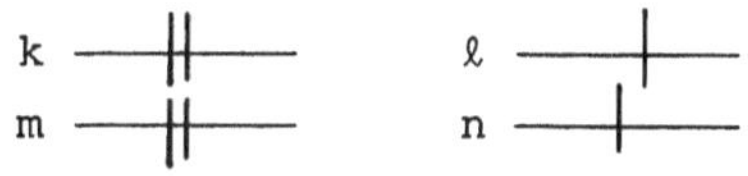

Triplet and singlet states

	ω	α	C	m	n	k	ℓ	β	C	m	n	k	ℓ
C	2		2	3/2	3/2	3/2	3/2		1	3/4	3/4	3/4	3/4
m	3/2		3/2	1	1	5/4	1		3/4	1/2	1/2	γ	1/2
n	3/2		3/2	1	1	1	5/4		3/4	1/2	1/2	1/2	γ
k	3/2		3/2	5/4	1	1	1		3/4	γ	1/2	1/2	1/2
ℓ	3/2		3/2	1	5/4	1	1		3/4	1/2	γ	1/2	1/2

For the triplet $\gamma = 3/4$

For the singlet $\gamma = 1/4$

B8.2

Triplet and singlet states

	ω	α	C	m	n	k	ℓ	β	C	m	n	k	ℓ
C	2		2	3/2	3/2	3/2	3/2		1	3/4	3/4	3/4	3/4
m	3/2		3/2	1	1	1	5/4		3/4	1/2	1/2	1/2	γ
n	3/2		3/2	1	1	5/4	1		3/4	1/2	1/2	γ	1/2
k	3/2		3/2	1	5/4	1	1		3/4	1/2	γ	1/2	1/2
ℓ	3/2		3/2	5/4	1	1	1		3/4	γ	1/2	1/2	1/2

For the triplet $\gamma = 3/4$
For the singlet $\gamma = 1/4$

C. Threefold degenerate open shell orbitals

C1. Three degenerate orbitals and one electron

$k, \ell, m \in O$ Doublet state

	ω	α	C	O	β	C	O
C	2		2	1/3		1	1/6
O	1/3		1/3	0		1/6	0

C2. Three degenerate orbitals and two electrons (one in each orbital)

Triplet and singlet states

	ω	α	C	O_1	O_2	β	C	O_1	O_2
C	2		2	2/3	2/3		1	1/3	1/3
O_1	2/3		2/3	0	1/6		1/3	0	γ
O_2	2/3		2/3	1/6	0		1/3	γ	0

For the triplet $\gamma = 1/6$
For the singlet $\gamma = -1/6$

C3. Three degenerate orbitals and three electrons (one in each orbital)

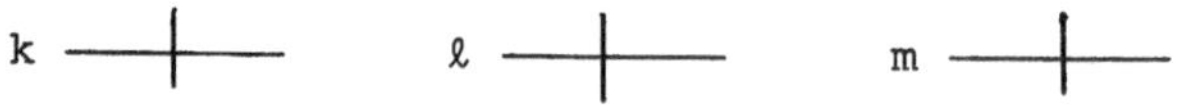

a Quadruplet and two doublet states
Same parameters as in A4.1 and A4.2

C4. Three degenerate orbitals and four electrons

Triplet and singlet states

	ω	α	C	O_1	O_2	β	C	O_1	O_2
C	2		2	4/3	4/3		1	2/3	2/3
O_1	4/3		4/3	2/3	5/6		2/3	1/3	γ
O_2	4/3		4/3	5/6	2/3		2/3	γ	1/3

For the triplet $\gamma = 1/2$
For the singlet $\gamma = 1/6$

C5. Three degenerate orbitals and five electrons

	ω	α	C	O_1	O_2	β	C	O_1	O_2
C	2		2	5/3	5/3		1	5/6	5/6
O_1	5/3		5/3	4/3	2/3		5/6	2/3	1/3
O_2	5/3		5/3	2/3	4/3		5/6	1/3	2/3

C6. A nondegenerate orbital (one electron) and three degenerate orbitals (one electron)

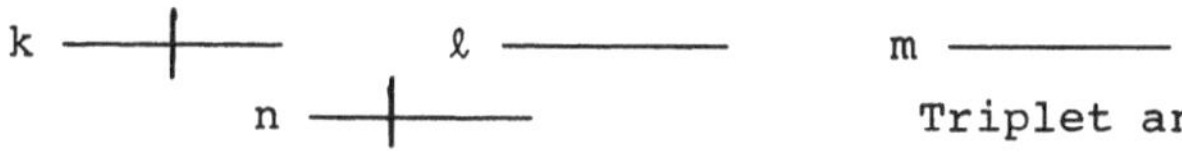

$k, \ell, m \in O$
Triplet and singlet states

	ω	α	C	n	O	β	C	n	O
C	2		2	1	1/3		1	1/2	1/6
n	1		1	0	1/6		1/2	0	γ
O	1/3		1/3	1/6	0		1/6	γ	0

For the triplet $\gamma = 1/6$
For the singlet $\gamma = -1/6$

C7. A nondegenerate orbital (one electron) and three degenerate orbitals (two electrons)

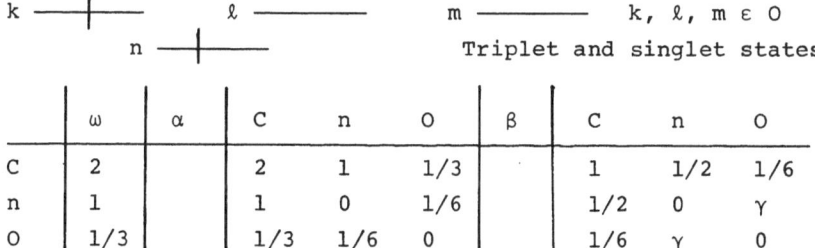

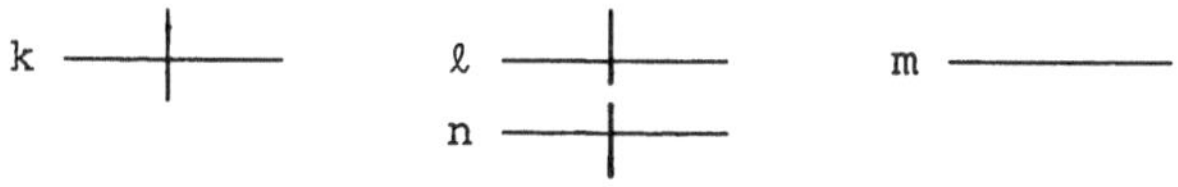

C7.1 Quadruplet

	ω	α	C	n	o_1	o_2	β	C	n	o_1	o_2
C	2		2	1	2/3	2/3		1	1/2	1/3	1/3
n	1		1	0	2/3	2/3		1/2	0	2/3	2/3
o_1	2/3		2/3	2/3	0	1/3		1/3	2/3	0	1/3
o_2	2/3		2/3	2/3	1/3	0		1/3	2/3	1/3	0

C7.2 Doublet 1

	ω	α	C	n	k	ℓ	m	β	C	n	k	ℓ	m
C	2		2	1	2/3	2/3	2/3		1	1/2	1/3	1/3	1/3
n	1		1	0	1/3	1/3	1/3		1/2	0	1/3	1/12	1/6
k	2/3		2/3	1/3	0	1/6	1/6		1/3	1/3	0	-1/12	-1/12
ℓ	2/3		2/3	1/3	1/6	0	1/6		1/3	1/12	-1/12	0	-1/12
m	2/3		2/3	1/3	1/6	1/6	0		1/3	1/6	-1/12	-1/12	0

C7.3 Doublet 2

	ω	α	C	n	o_1	o_2	β	C	n	o_1	o_2
C	2		2	1	2/3	2/3		1	1/2	1/3	1/3
n	1		1	0	1/3	1/3		1/2	0	1/6	1/6
o_1	2/3		2/3	1/3	0	1/6		1/3	1/6	0	-1/6
o_2	2/3		2/3	1/3	1/6	0		1/3	1/6	-1/6	0

Appendix B

Slater Rules

The matrix elements of an electronic hamiltonian operator $\mathcal{H}$ in the basis of monodeterminantal Slater functions are used in some parts of this work, and for this reason the corresponding expressions for each matrix element are given here.
Let's take Ψ_I and Ψ_J as two such functions, and $\{\Theta_i\}$ as a chosen set of spinorbitals. The normalized Slater functions will be represented as

$$\Psi = |\Theta_1\Theta_2 \ldots\Theta_n|.$$

The functions Ψ_I and Ψ_J will be considered as already having the maximal coincidence in the ordering of the string of spinorbitals inside each function. That is, if

$$\Psi_I = |\Theta_1^I\Theta_2^I \ldots \Theta_n^I|$$

and

$$\Psi_J = |\Theta_1^J\Theta_2^J \ldots \Theta_n^J|,$$

expressing $N_i = \delta(\Theta_i^I = \Theta_i^J)$ as a measure of the coincidence of spinorbitals, Θ_i^I and Θ_i^J being at the same place in both functions, then in order to use Slater rules the sum:

$$S = \sum_{i=1}^{n} N_i$$

shall be maximal. If this is not so, a reordering of the spinorbitals inside the determinants should take place, remembering that:

$$\Psi = |\Theta_1\Theta_2 \ldots \Theta_p \ldots \Theta_q \ldots \Theta_n| = - |\Theta_1\Theta_2 \ldots \Theta_q \ldots \Theta_p \ldots \Theta_n|,$$

as easily follows from the properties of determinants.
Supposing we have attained the situation of max[S], Slater rules can be given through four well defined occurrences:

1) $\Psi_I = \Psi_J$, that is max[S] = n:

$$\langle\Psi_I|\mathcal{H}|\Psi_I\rangle = \sum_i (\langle\Theta_i|h|\Theta_i\rangle + \sum_{j>i} [(\Theta_i\Theta_i|\Theta_j\Theta_j) - (\Theta_i\Theta_j|\Theta_i\Theta_j)].$$

2) Ψ_I and Ψ_J have one different spinorbital, or max[S] = n - 1. Taking Θ_i as the Ψ_I spinorbital in question and Θ_j as the corresponding Ψ_J one, then:

$$<\Psi_I|\mathcal{H}|\Psi_J> = <\Theta_i|h|\Theta_j> + \sum_{k\neq i,j} [(\Theta_k\Theta_k|\Theta_i\Theta_j) - (\Theta_k\Theta_i|\Theta_k\Theta_j)].$$

3) Ψ_I and Ψ_J have two different spinorbitals, or max[S] = n - 2. Taking as different orbitals Θ_p^I and Θ_q^I in Ψ_I and Θ_r^J, Θ_s^J in Ψ_J, in this case

$$<\Psi_I|\mathcal{H}|\Psi_J> = (\Theta_p^I\Theta_r^J|\Theta_q^I\Theta_s^J) - (\Theta_p^I\Theta_s^J|\Theta_q^I\Theta_r^J).$$

4) Ψ_I and Ψ_J have more than two different spinorbitals or max[S] $\leq$ n - 3, we have for all these situations

$$<\Psi_I|\mathcal{H}|\Psi_J> = 0$$

In the previous expressions h is the monoelectronic hamiltonian and

a) $$(\Theta_i|h|\Theta_j) = \int \Theta_i^*(1)\ h(1)\ \Theta_j(1)\ d\zeta(1)$$

b) $$(\Theta_i\Theta_j|\Theta_k\Theta_\ell) = \iint \Theta_i^*(1)\ \Theta_k^*(2)\ \frac{1}{r_{12}}\ \Theta_j(1)\ \Theta_\ell(2)\ d\zeta(1)\ d\zeta(2),$$

where the volume elements $d\zeta$ contain space and spin coordinates. So if we express each spinorbital Θ as a product of space, Φ, and spin, σ, parts:

$$\Theta = \Phi\ \sigma,$$

then

a) $$(\Theta_i|h|\Theta_j) = (\Phi_i|h|\Phi_j)\ \delta(\sigma_i = \sigma_j)$$

$$= h_{ij}\ \delta(\sigma_i = \sigma_j)$$

b) $$(\Theta_i\Theta_j|\Theta_k\Theta_\ell) = (\Phi_i\Phi_j|\Phi_k\Phi_\ell)\ \delta(\sigma_i = \sigma_j)\ \delta(\sigma_k = \sigma_\ell)$$

$$= (ij|k\ell)\ \delta(\sigma_i = \sigma_j)\ \delta(\sigma_k = \sigma_\ell).$$

These expressions can be used now to simplify the hamiltonian elements and transform them into spatial orbital expressions, for each particular case. Using logical Kronecker delta symbols:

1) $\max[S] = n$

$$\langle\Psi_I|\mathscr{H}|\Psi_I\rangle = \sum_i \omega_i h_{ii} + \sum_i \sum_j (\alpha_{ij}J_{ij} - \beta_{ij}K_{ij})$$

with

$$\omega_i = \delta(\sigma_i = \alpha) + \delta(\sigma_i = \beta)$$

$$\alpha_{ij} = \frac{1}{2}\{\delta(\sigma_i,\sigma_j = \alpha) + \delta(\sigma_i,\sigma_j = \beta)$$

$$+ \delta(\sigma_i = \alpha,\ \sigma_j = \beta) + \delta(\sigma_i = \beta,\ \sigma_j = \alpha)\}$$

$$\beta_{ij} = \frac{1}{2}\ \{\delta(\sigma_i,\sigma_j = \alpha) + \delta(\sigma_i,\sigma_j = \beta)\}$$

and using $J_{ii} = K_{ii}$

2) $\max[S] = n - 1$

$$\langle\Psi_I|\mathscr{H}|\Psi_J\rangle = \delta(\sigma_i = \sigma_j)h_{ij}$$

$$+ \sum_{k\neq i,j} \{\delta(\sigma_i = \sigma_j)(kk|ij) - \delta(\sigma_k = \sigma_i)\ \delta(\sigma_k = \sigma_j)(ki|kj)\}$$

3) $\max[S] = n - 2$

$$\langle\Psi_I|\mathscr{H}|\Psi_J\rangle = \delta(\sigma_p = \sigma_r)\ \delta(\sigma_q = \sigma_s)(pr|qs)$$

$$- \delta(\sigma_p = \sigma_s)\ \delta(\sigma_q = \sigma_r)(ps|qr).$$

Appendix C

Multiconfigurational Fock Operators

We will give here the formalism of Fock operators for some multiconfigurational cases, as an intermediate tool to be used in the multiconfigurational coupling operator scheme. The Fock operators which are given here are related with singlet, doublet and triplet states, and can cover the most usual situations.

1. Multiconfigurational SCF Theory of Double and Single Excitations on a Closed Shell Groundstate

Consider a system with a set of doubly occupied MO's $C = \{|i\rangle\}$ and a virtual set $V = \{|v\rangle\}$.
A single excitation wavefunction corresponding to the promotion of an electron from the doubly occupied orbital $|i\rangle$ to the vacant orbital $|v\rangle$ will be represented by $\Psi(i,v)$. Similarly, a paired excitation function, attached to the promotion of two electrons from $|i\rangle$ to $|v\rangle$, will be given by $\Psi(ii,vv)$. The total wavefunction can be constructed as

$$\Psi = S_o\Psi_o + \sum_i \sum_v (S_{iv}\Psi(i,v) + D_{iv}\Psi(ii,vv)) \qquad (1.C)$$

Ψ_o being the ground state wavefunctions and S_o, $\{S_{iv}\}$ and $\{D_{iv}\}$ the variational coefficients. The various matrix elements of the Configuration Interaction matrix can be expressed as

$$\langle\Psi o|\mathcal{H}|\Psi o\rangle = \sum_i (\varepsilon_{ii} + E_{ii}) + E_o \qquad (2a.C)$$

$$\langle\Psi(i,u)|\mathcal{H}|\Psi(j,v)\rangle = \delta_{ij}\delta_{uv}E_o + \delta_{ij}E_{uv} \qquad (2b.C)$$

$$- \delta_{uv}E_{ij} + 2(iu|jv) - (ij|uv)$$

$$\langle\Psi(ii,uu)|\mathcal{H}|\Psi(jj,vv)\rangle = \qquad (2c.C)$$

$$\delta_{ij}\delta_{uv}\,[E_o + 2\,E_{uu} - 2\,E_{ii} - 2(2\,J_{iu} - K_{iu})]$$

$$+ \delta_{ij}K_{uv} + \delta_{uv}K_{ij}$$

$$<\Psi_o|\mathcal{H}|\Psi(i,u)> = (2)^{-1/2}\ E_{iu} \tag{2d.C}$$

$$<\Psi_o|\mathcal{H}|\Psi(ii,uu)> = K_{iu} \tag{2e.C}$$

$$<\Psi(i,u)|\mathcal{H}|\Psi(jj,vv)> = \tag{2f.C}$$

$$(2)^{-1/2}\ [\delta_{ij}\delta_{uv}E_{iu} + \delta_{ij}(iv|uv) - \delta_{uv}(ij|uj)]$$

with the aid of:

* $\varepsilon_{pq} = <p|h|q>$

* $E_{pq} = \varepsilon_{pq} + \sum_{j\varepsilon C} [2(jj|pq) - (jp|jq)]$

$= <p|F_c|q>$

$= E_{qp}$

* $F_c = h = \sum_j (2\ J_j - K_j).$

Taking into account the matrix elements obtained before, the expression of the energy given by a wavefunction like (1.C) is (i, jεC; u, vεV):

$$E = <\Psi|\mathcal{H}|\Psi> \tag{3.C}$$

$$= E_o + \sum_{u,v} A_{uv}E_{uv} + \sum_{i,u} A_{iu}E_{iu}$$

$$+ \sum_{i,j} A_{ij}E_{ij} + \sum_{u,v} B_{uv}K_{ub} + \sum_{i,j} B_{ij}K_{ij}$$

$$+ \sum_i \sum_{u,v} B^v_{iu}(iv|vu) + \sum_{i,j} \sum_u B^j_{iu}(ij|uj)$$

$$+ \sum_{i,j} \sum_{u,v} [C^{uv}_{ij}(iu|jv) - D^{uv}_{ij}(ij|uv)]$$

with the parameters

* $A_{uv} = A_{vu} = \sum_i (S_{iu}S_{iv} + 2\ \delta_{uv}D^2_{iu})$

$$* \quad A_{iu} = 2(2)^{-1/2} \; (S_o S_{iu} + S_{iu} D_{iu})$$

$$* \quad A_{ij} = - \sum_u (S_{iu} S_{ju} + \delta_{ij} D_{iu}^2) = A_{ji}$$

$$* \quad B_{uv} = B_{vu} = \sum_i D_{iu} D_{iv}$$

$$* \quad B_{ij} = B_{ji} = \sum_u D_{iu} D_{ju}$$

$$* \quad B_{iu}^{v} = 2(2)^{-1/2} \; S_{iu} D_{iv}$$

$$* \quad B_{iu}^{j} = -2(2)^{-1/2} \; S_{iu} D_{ju}$$

$$* \quad C_{ij}^{uv} = C_{ji}^{vu} - 2(S_{iu} S_{jv} + \delta_{ij} \delta_{uv} (D_{iu}^2 + S_o D_{iu}))$$

$$* \quad D_{ij}^{uv} = D_{ji}^{vu} = S_{iu} S_{jv} + 4 \; \delta_{ij} \delta_{uv} D_{iu}^2 .$$

By variation of the energy (3.C) and then rearranging some terms, the following Fock operators are found:

$$F_{ij} = \omega_{ij} F_c + \sum_{u\varepsilon V} \sum_{v\varepsilon V} (\alpha_{ij}^{uv} \, V_{uv} - \beta_{ij}^{uv} \, X_{uv}) \tag{4a.C}$$

$$+ \; \delta_{ij} \sum_{k\varepsilon C} \sum_{u\varepsilon V} (\alpha_{ku} \, V_{ku} - \beta_{ku}^{i} \, X_{ku})$$

$$+ \; \delta_{ij} \sum_{k\varepsilon C} \sum_{\ell\varepsilon C} (\alpha_{k\ell} \, V_{k\ell} - \beta_{k\ell}^{i} \, X_{k\ell})$$

$$F_{iu} = \omega_{iu} F_c + \sum_{v \leftarrow V} \alpha_{iu}^{v} \, K_v + \sum_{j\varepsilon C} \alpha_{iu}^{j} \, K_j \tag{4b.C}$$

$$F_{uv} = \omega_{uv} F_c + \delta_{uv} \; \{ \sum_{X\varepsilon V} \alpha_{ik} \, K_X \tag{4c.C}$$

$$+ \sum_{i\varepsilon C} \sum_{X\varepsilon V} \alpha_{iX}^{u} \, X_{iX} + \sum_{i\varepsilon C} \sum_{j\varepsilon C} (\alpha_{uv}^{ij} \, V_{ji} \; \beta_{uv}^{ij} \, X_{ji})$$

where

$$* \quad \omega_{ij} = 2 \; \delta_{ij} + A_{ij} = \omega_{ji}$$

* $\alpha_{ij}^{uv} = 2\ \delta_{ij} A_{uv} - D_{ij}^{uv} = \alpha_{ji}^{vu}$

* $\beta_{ij}^{uv} = \delta_{ij} A_{uv} - C_{ij}^{uv} = \beta_{ji}^{vu}$

* $\alpha_{ku} = 2\ A_{ku}$

* $\beta_{ku}^{i} = A_{ku} - B_{ku}^{i}$

* $\alpha_{k\ell} = 2\ A_{k\ell} = \alpha_{\ell k}$

* $\beta_{k\ell}^{i} = A_{k\ell} - \delta_{k\ell}\ B_{ik} = \beta_{\ell k}^{i}$

* $\omega_{iu} = A_{iu}$

* $\alpha_{iu}^{v} = B_{iu}^{v}$

* $\alpha_{iu}^{j} = B_{iu}^{j}$

* $\omega_{uv} = A_{uv} = \omega_{vu}$

* $\alpha_{ux} = 2\ B_{ux} = \alpha_{xu}$

* $\alpha_{ix}^{u} = B_{ix}^{u}$

* $\alpha_{uv}^{ij} = -\ D_{ij}^{uv} = \alpha_{vu}^{ij}$

* $\beta_{uv}^{ij} = -\ C_{ij}^{uv} = \beta_{vu}^{ji}$.

As special cases, if only monoexcitations are taken into account all the elements of the set $\{D_{iu}\}$ are null, or if only paired excitations are considered the set $\{S_{iu}\}$ is not considered. The expressions obtained in this latter case coincide then with those of section (2.V). The total energy can be expressed through

$$E = \frac{1}{2} \{\sum_i \sum_j \langle i|F_{ij} + \omega_{ij}\ h|j\rangle$$

$$+ \sum_{i}\sum_{u} \langle i|F_{iu} = \omega_{iu}\, h|u\rangle$$

$$+ \sum_{u}\sum_{v} \langle u|F_{uv} + \omega_{uv}\, h|v\rangle\}.$$

2. Multiconfigurational SCF Theory for Doublet States

In order to construct the multiconfigurational wavefunction for this case we shall consider that there are n + m + 1 MO's, where n are doubly occupied, m are void and one singly occupied. Any set of orbital indexes having this occupation pattern will be named S_k. The number of sets S_k is $\binom{n+m+1}{n}$.

If we denote the orbital index filled with one electron by I the total wavefunction can be written as

$$\Psi = \sum_{I}\sum_{k} C_k^I\, \Psi_I(S_k) \tag{5.C}$$

with

$$C_k^I = 0,\ (\forall I,k|I\varepsilon S_k).$$

In the energy expression attached to (5.C) the terms

$$T = \langle\Psi_I(S_k)|\mathcal{H}|\Psi_J(S_\ell)\rangle$$

can be calculated according to the following cases:

* a) $\quad I = J,\ S_k = S_\ell;\ \#(S_k \cap S_\ell) = n.$

The wavefunction will be represented by

$$\Psi_I(S_k) = |A\bar{A}\ \ldots\ P\bar{P}I|$$

and

$$T_a = \delta_{IJ}\delta_{k\ell}\{2\sum_{N} h_{NN} + \sum_{M,N}(2\, J_{MN} - K_{MN})$$

$$+ h_{II} + \sum_{M}(2\, J_{IM} - K_{IM})\}\ \delta[M,N\varepsilon(S_k \cap S_\ell)]$$

* b) $I = J,\ S_k \neq S_\ell;\ \#(S_k \cap S_\ell) = n - 1$

$$T_b = \delta_{IJ}\ \delta[n-1 = \#(S_k \cap S_\ell);\ M\varepsilon S_k;\ N\varepsilon S_\ell;\ M,N\not\varepsilon (S_k \cap S_\ell)]\ K_{MN}$$

* c) $I \neq J;\ S_k = S_\ell$

$$T_c = (1 - \delta_{IJ})\ \delta_{k\ell}\delta[M\varepsilon(S_k \cap S_\ell)]\{h_{IJ} + \sum_M [2(IJ|MM) - (IM|JM)]\}$$

* d) $I \neq J;\ S_k \neq S_\ell;\ \#(S_k \cap S_\ell) = n - 1$

** d1) $I\varepsilon S_\ell;\ J\varepsilon S_k$.

The wavefunctions attached to this case are

$$\Psi_I(S_k) = |\ldots J\bar{J}I| = -\ |\ldots JI\bar{J}|$$

and

$$\Psi_J(S_\ell) = |\ldots I\bar{I}J| = |\ldots JI\bar{I}|$$

so the hamiltonian term can be written as

$$T_{d_1} = (1 - \delta_{IJ})\ \delta[n-1 = \#(S_k \cap S_\ell);\ I\varepsilon S_\ell;\ J\varepsilon S_k;\ M\varepsilon(S_k \cap S_\ell)]$$

$$x(-1)\{h_{IJ} + (II|IJ) + (JJ|JI) + \sum_M [2(IJ|MM) - (IM|JM)]\}$$

** d2) $I = S_\ell;\ J \not\varepsilon S_k$.

The wavefunctions can be symbolized by

$$\Psi_I(S_k) = |\ldots M\bar{M}I|$$

and

$$\Psi_J(S_\ell) = |\ldots I\bar{I}J|$$

with

$$T_{d_2} = (1 - \delta_{IJ})\ \delta[n-1 = \#(S_k \cap S_\ell);\ I\varepsilon S_\ell;\ J\varepsilon S_k]$$

$$\times\ \delta[M\varepsilon S_k;\ M\not\varepsilon(S_k \cap S_\ell\ )]\ \ (-1)(IM|JM)$$

**d3) $I\not\varepsilon S_\ell,\ J\varepsilon S_k.$

In this case

$$\Psi_I(S_k) = |\ldots\ J\bar{J}I|$$

and

$$\Psi_J(S_\ell) = |\ldots\ M\bar{M}J|.$$

So

$$T_{d_3} = \delta[n-1 = \#(S_k \cap S_\ell);\ I\not\varepsilon S_\ell;\ J\not\varepsilon S_k]\ \ (1 - \delta_{IJ})$$

$$\times\ \delta[M\varepsilon S_\ell;\ M\not\varepsilon(S_k \cap S_\ell)](-1)(IM|JM).$$

Using these results the total energy can be written as

$$E = \sum_I \sum_J \omega_{IJ}\ H_{IJ} + \sum_I \sum_J \sum_M \{\alpha^M_{IJ}(IJ|MM) - \beta^M_{IJ}(IM|JM)\}$$

with

a)
$$\omega_{IJ} = \omega_{JI} = 2\ \delta_{IJ} \sum_k \sum_N (C^N_k)^2\ \delta(I\varepsilon S_k) + \sum_k C^I_k\ C^J_k$$

$$- \sum_k \sum_\ell (1-\delta_{IJ}) C^I_k\ C^J_\ell\ \delta[n-1 = \#(S_k \cap S_\ell);\ I\varepsilon S_\ell;\ J\varepsilon S_k]$$

b)
$$\alpha^M_{IJ} = \alpha^M_{JI} = 2\ \delta_{IJ} \sum_k \sum_N (C^N_k)^2\ \delta(I,M\varepsilon S_k) + 2 \sum_k C^I_k\ C^J_k\ \delta(M\varepsilon S_k)$$

$$- \sum_k \sum_\ell C^I_k\ C^I_\ell(1 - \delta_{IJ})\ \delta[n-1 = \#(S_k \cap S_\ell);\ I\varepsilon S_\ell;\ J\varepsilon S_k]$$

$$x\ (\delta_{IM} + \delta_{JM} + 2\ \delta[M\varepsilon(S_k \cap S_\ell)])$$

c) $$\beta^M_{IJ} = \beta^M_{JI} = \delta_{IJ} \sum_k \sum_N (C^N_k)^2\ \delta(I, M\varepsilon S_k) + \sum_k C^I_k\ C^J_k\ \delta(M\varepsilon S_k)$$

$$- \delta_{IJ} \sum_k \sum_\ell \sum_N C^N_k\ C^N_\ell\ \delta[n-1 = \#(S_k \cap S_\ell);\ M\varepsilon S_k;\ I\varepsilon S_\ell;\ I,M\not\varepsilon(S_k \cap S_\ell)]$$

$$- \sum_k \sum_\ell C^I_k\ C^J_\ell (1-\delta_{IJ})\ \delta[n-1 = \#(S_k \cap S_\ell)]\ \{\delta[I\varepsilon S_\ell;\ J\varepsilon S_k;\ M\varepsilon(S_k \cap S_\ell)]$$

$$- \delta[M\not\varepsilon(S_k \cap S_\ell)][\delta(I\varepsilon S_\ell;\ J\not\varepsilon S_k;\ M\varepsilon S_k) + \delta(I\not\varepsilon S_\ell;\ J\varepsilon S_k;\ M\varepsilon S_\ell)]\}$$

and the Fock operators found are

$$F_{IJ} = \omega_{IJ} h + \sum_M \sum_L (\alpha^{ML}_{IJ}\ V_{ML} - \beta^{ML}_{IJ}\ X_{ML})$$

where

$$\alpha^{ML}_{IJ} = \delta_{ML}\ \alpha^M_{IJ} + \delta_{IJ}\ \alpha^I_{ML}$$

and

$$\alpha^{ML}_{IJ} = \delta_{ML}\ \beta^M_{IJ} + \delta_{IJ}\ \beta^I_{ML} .$$

It is easy to deduce

$$F^*_{IJ} = F_{JI}$$

and also that

$$E = \frac{1}{2} \sum_I \sum_J \langle I|F_{IJ} + \omega_{IJ} h|J\rangle.$$

3. Multiconfigurational SCF Theory for Triplet States

Starting from a closed shell groundstate one has two orbitals: the doubly occupied $\mathscr{C}$ and the virtual $\mathscr{V}$. In the triplet MC we consider the mono excitations from an orbital $I \varepsilon \mathscr{C}$ to $V \varepsilon \mathscr{V}$, and in some instances also the excitations from two electrons of $J \varepsilon \mathscr{C}$ to $X \varepsilon \mathscr{V}$.

On these grounds one can write the total wavefunction as

$$\Psi = \sum_{I}\sum_{V} T_{IV}\ \Psi(I,V) + \sum_{I}\sum_{J}\sum_{V}\sum_{X} T_{IV}^{JX}\ \Psi(I,V;J,X)$$

where $\Psi(I,V)$ represents a monoexcitation from I to V and $\Psi(I,V;J,X)$ a monoexcitation of the latter type given above followed by the paired excitation of two electrons of J to X.
The indexes $\{I,\bar{I},J,\bar{J},K,L, \ldots\}$ are supposed to be assigned to orbitals of $\mathscr{C}$ and $\{V,\bar{V},X,\bar{X},Y,Z, \ldots\}$ to orbitals of $\mathscr{V}$. The total energy can be divided into four terms,

$$E = E_{SS} + (E_{SD} + E_{DS}) + E_{DD}$$

where the single excitation, single and paired, and the paired interactions are collected in each term.
A tedious analysis of all the possible cases, following the same general trends as in the cases studied before gives:

$$E = E_0 + \sum_{X}\sum_{V} \gamma_{XV}\ E_{XV} + \sum_{I}\sum_{J} \gamma_{IJ}\ E_{IJ}$$

$$+ \sum_{I}\sum_{J}\sum_{V}\sum_{X} \{\alpha_{IJ}^{XV}(IJ|VX) - \beta_{IJ}^{XV}(IX|JV)\}$$

$$+ \sum_{X}\sum_{V}\sum_{U} \{\alpha_{U}^{XV}(UU|XV) - \beta_{U}^{XV}(UV|UX)\}$$

$$+ \sum_{I}\sum_{J}\sum_{K} \{\alpha_{K}^{IJ}(IJ|KK) - \beta_{K}^{IJ}(IK|JK)\}$$

with the state parameters γ, α and β depending on a complicated manner of the coefficients T, and the term

$$E_0 = \sum_{I} \langle I|2h + \sum_{K} (2\ J_K - K_K)|I\rangle$$

being the contribution of the closed shell part.
Variation of the total energy gives the Fock operators

$$F_{IJ} = \omega_{IJ}F_C + \sum_{K}\sum_{L} (\mu_{IJ}^{KL}\ V_{KL} - \nu_{IJ}^{KL}\ X_{KL})$$

$$+ \sum_X \sum_V (\mu_{IJ}^{XV} V_{VX} - \nu_{IJ}^{XV} X_{VX})$$

and

$$F_{XV} = \gamma_{XV} F_C + \sum_I \sum_J (\alpha_{IJ}^{XV} V_{JI} - \beta_{IJ}^{XV} X_{JI})$$

$$+ \sum_U \sum_Z (\mu_{UZ}^{XV} V_{UZ} - \nu_{UZ}^{XV} X_{UZ})$$

with

a) $$\omega_{IJ} = 2\,\delta_{IJ} + \gamma_{IJ} = \omega_{JI}$$

$$= 2\,\delta_{IJ}[1 - \sum_M \sum_V \sum_X (T_{MV}^{JX})^2] - \sum_Z T_{IZ} T_{JZ}$$

$$+ \sum_V \sum_X (T_{IV}^{JX}\, T_{JV}^{IX} - \sum_M T_{IV}^{MX}\, T_{JV}^{MX})$$

b1) $$\mu_{IJ}^{KL} = 2\,\delta_{IJ}\gamma_{KL} + \delta_{IJ}\,\alpha_I^{KL} + \delta_{KL}\,\alpha_K^{IJ} = \mu_{JI}^{LK}$$

$$= 2\,\delta_{IJ}\,\{\sum_V \sum_X (T_{KV}^{LX}\, T_{LV}^{KX} + T_{KV}^{IX}\, T_{LV}^{JX})$$

$$- \sum_M \sum_V \sum_X [2\,\delta_{KL}\,(T_{MV}^{KX})^2 + T_{KV}^{MX}\, T_{LV}^{MX}]$$

$$- \sum_Z T_{KZ}\, T_{LZ}\} + 2\,\delta_{KL} \sum_V \sum_X T_{IV}^{KX}\, T_{JV}^{KX}$$

b2) $$\mu_{IJ}^{XV} = 2\,\delta_{IJ}\,\gamma_{XV} + \alpha_{IJ}^{XV} = \mu_{JI}^{VX}$$

$$\mu_{UZ}^{XV} = \delta_{XV}\,\alpha_X^{UZ} + \delta_{UX}\,\alpha_U^{XV} = \mu_{ZU}^{VX}$$

$$= 2 \sum_I \sum_J (\,\delta_{XV}\, T_{IZ}^{JX}\, T_{IU}^{JX} + \delta_{UZ}\, T_{IV}^{JU}\, T_{IX}^{JU})$$

c1) $$\nu_{IJ}^{KL} = \delta_{IJ}(\gamma_{KL} + \beta_I^{KL}) + \delta_{KL}\,\beta_K^{IJ} = \nu_{JI}^{LK}$$

$$= - \delta_{IJ} \sum_{Z} T_{KZ} T_{LZ}$$

$$+ \sum_{X} \sum_{V} \{\delta_{KL} [T^{KX}_{IV} T^{KX}_{JV} + T^{JX}_{IV} T^{KX}_{JV} + T^{KX}_{IV} T^{IX}_{JV}]$$

$$+ \delta_{IJ} [T^{LX}_{KV} T^{KX}_{LV} + T^{IX}_{KV} T^{IX}_{LV} + T^{LX}_{KV} T^{IX}_{LV} + T^{IX}_{KV} T^{KX}_{LV}]$$

$$- \sum_{M} \sum_{V} \sum_{X} \{\delta_{IJ} \delta_{KL} [2 (T^{KX}_{MV})^2 + T^{JX}_{MV} T^{KX}_{MV} + T^{LX}_{MV} T^{IX}_{MV}]$$

$$+ \delta_{IJ} T^{MX}_{KV} T^{MX}_{LV}\}$$

c2) $$\nu^{XV}_{IJ} = \delta_{IJ} \gamma_{XV} + \beta^{XV}_{IJ} = \nu^{VX}_{JI}$$

c3) $$\nu^{XV}_{UZ} = \delta_{XV} \beta^{UZ}_{X} + \delta_{UZ} \beta^{XV}_{U} = \nu^{VX}_{ZU}$$

$$= \delta_{XV} \sum_{I} \sum_{J} (T^{JX}_{IZ} T^{JX}_{IU} + T^{JU}_{IZ} T^{JX}_{IU} + T^{JX}_{IZ} T^{JX}_{IU})$$

$$+ \delta_{UZ} \sum_{I} \sum_{J} (T^{JU}_{IV} T^{JU}_{IX} + T^{JX}_{IV} T^{JU}_{IX} + T^{JU}_{IV} T^{JV}_{IX})$$

$$- 2 \delta_{XV} \delta_{UZ} \sum_{I} \sum_{J} \sum_{Y} T^{JX}_{IY} T^{JU}_{IY}$$

d) $$\gamma_{XV} = \sum_{K} T_{KV} T_{KX} - \sum_{K} \sum_{L} T^{LX}_{KV} T^{LV}_{KX}$$

$$+ \sum_{K} \sum_{L} \sum_{Z} [2 \delta_{XV} (T^{LX}_{KZ})^2 + T^{LZ}_{KV} T^{LZ}_{KX}]$$

e) $$\alpha^{XV}_{IJ} = - T_{IV} T_{JX} - T^{JX}_{IV} T^{IV}_{JX}$$

$$+ \sum_{K} (T^{KX}_{IV} T^{KV}_{JX} + 2 \delta_{IJ} T^{JX}_{KV} T^{JV}_{KX})$$

$$+ \sum_{Z} (T^{JX}_{IV} T^{IZ}_{JX} + 2 \delta_{XV} T^{JX}_{JZ} T^{IX}_{JZ})$$

$$- \sum_{K} \sum_{Z} [T^{KZ}_{IV} T^{KZ}_{JX} + 2 \delta_{IJ} T^{JZ}_{KV} T^{JZ}_{KX}$$

$$+ 2\,\delta_{XV}\, T^{KX}_{IZ}\, T^{KX}_{JZ} + 4\,\delta_{IJ}\,\delta_{XV}(T^{JX}_{KZ})^2]$$

f) $$\beta^{VX}_{IJ} = - T_{JX}\, T^{JX}_{IV} - T_{IV}\, T^{IV}_{JX}$$

$$+ \sum_K [\delta_{IJ}(T_{KX}T^{JX}_{KV} + T_{KV}T^{JV}_{KX} + T^{JX}_{KV}T^{JV}_{KX}) + T^{KX}_{IV}T^{KV}_{JX}]$$

$$+ \sum_Z [\delta_{KV}(T_{JZ}T^{JX}_{IZ} + T_{IZ}T^{IX}_{JZ} + T^{JX}_{JZ}T^{IX}_{JZ}) + T^{JZ}_{IV}T^{IZ}_{JX}]$$

$$- \sum_K \sum_Z \{\delta_{IJ}T^{JZ}_{KV}T^{JZ}_{KX} + \delta_{XV}T^{KX}_{IZ}T^{KX}_{JZ}$$

$$+ 2\,\delta_{IJ}\,\delta_{XV}[T_{KZ}T^{JX}_{KZ} + (T^{JX}_{KZ})^2]\}.$$

It is easy to check that

$$F^*_{IJ} = F_{JI}$$

and

$$F^*_{XV} = F_{VX}\ .$$

The energy can also be expressed as:

$$E = \frac{1}{2}\{\sum_I \sum_J \langle I|F_{IJ} + \omega_{IJ}h|J\rangle$$

$$+ \sum_X \sum_V \langle X|F_{XV} + \gamma_{XV}h|V\rangle\}\ .$$

Suggested Reading

Chapter II

*K. Hirao, H. Nakatsuji; J. Chem. Phys., 59, 1457 (1973).
*K. Hirao; J. Chem. Phys., 60, 3215 (1974).
*R. Caballol, R. Gallifa, J.M. Riera, R. Carbo; Intl. J. Quantum Chem., 8, 373 (1974).
*R. Carbo, R. Gallifa, J.M. Riera; Chem. Phys. Lett., 30, 43 (1975).

Chapter III

a) Eigenspace Manipulation

*S. Huzinaga, C. Arnau; Phys. Rev. A, 1, 1285 (1970).
*S. Huzinaga, C. Arnau; J. Chem. Phys., 54, 1948 (1971).
*C. Arnau, R. Carbo, S. Huzinaga; Intl. J. Quantum Chem., 6, 843 (1972).
*S. Huzinaga, D. McWilliams, A.A. Cantu; Adv. Quantum Chem., 7, 187 (1973).
*R. Carbo; Intl. J. Quantum Chem., 8, 423 (1974).

b) Convergence in SCF Theory

*D.H. Sleeman; Theoret. Chim. Acta, 11, 135 (1968).

c) Level Shift

*V.R. Saunders, I.H. Hillier; Intl. J. Quantum Chem., 7, 699 (1973).
*M.H. Wood, A. Veillard; Mol. Phys., 26, 595 (1973).
*R. Carbo, J.A. Hernandez, F. Sanz; Chem. Phys. Lett., 47, 581 (1977).

Chapter IV

*R. Carbo, R. Gallifa, J.M. Riera; Chem. Phys. Lett., 33, 545 (1975).

Chapter V

*E. Clementi, A. Veillard; J. Chem. Phys., 44, 3050 (1966).
*A. Veillard, E. Clementi; Theoret. Chim. Acta, 7, 133 (1967).
*G. Das, A.C. Wahl; J. Chem. Phys. 56, 1769 (1972).
*R. Carbo, J.A. Hernandez; Chem. Phys. Lett., 47, 85 (1977).

Chapter VI

a) Perturbation Theory

*J.O. Hirschfelder; Intl. J. Quantum Chem., 3, 731 (1969).

b) SCF Perturbation Theory

*R.T. Amos, G.G. Hall; Theoret. Chim. Acta, 5, 148 (1966).
*J. Bacon, D.P. Santry; J. Chem. Phys., 56, 2011 (1972).

c) Interaction

*J.N. Murrell; "Orbital Theories of Molecules and Solids", Ed. by N.H. March. Clarendon Press, Oxford (1974) p. 311.

Chapter VIII

a) Non-empirical approximation

*R.S. Mulliken, J. Chim. Phys., 46, 500 (1949).
*R.S. Mulliken, J. Chem. Phys., 23, 1833, 1841, 2338, 2343 (1955).
*K. Ruedenberg; J. Chem. Phys., 19, 1433 (1951).

b) Decomposition of Repulsion Matrix

*N.H.F. Beebe, J. Linderberg; Intl. J. Quantum Chem., (to be published).

c) Empirical Approximation

*J.A. Pople, D.L. Beveridge; "Approximate MO Theory". McGraw-Hill, New York (1970).

d) Model Potential

*D. McWilliams, S. Huzinaga; J. Chem. Phys., 63, 4678 (1975).

Chapter IX

a) Concept of Shell

*S. Huzinaga; J. Chem. Phys., 51, 3971 (1969).

b) Generalized Brillouin Theorem

*H. Hinze; J. Chem. Phys., 59, 6424 (1973).

c) Structure of SCF

*R. McWeeny; "M.O. in Chemistry, Physics and Biology". Eds. P.O. Löwdin, B. Pullman. Academic Press, New York (1964).
*B.T. Sutcliffe; Theoret. Chim. Acta, 33, 201 (1974); 39, 93 (1975).

d) Optimization of Non-linear Parameters

*P.W. Payne; J. Chem. Phys., 65, 1920 (1976).

Bibliographical Survey on Open Shell SCF Theory

1957

R. Lefebvre; J. Chim. Phys., 54, 168 (1957)
"L'Intéraction de Configurations comme méthode de Calcul des O.M. du Champ Self-Consistent. Etat fondamental d'un système à un nombre impair d'électrons"

1960

S. Huzinaga; Phys. Rev., 120, 866 (1960)
"Applicability of Roothaan's SCF Theory"

C.C.J. Roothaan; Rev. Mod. Phys., 32, 179 (1960)
"SCF Theory for Open Shells of Electronic Systems"

1961

S. Huzinaga; Phys. Rev., 122, 131 (1961)
"Analytical Methods in Hartree-Fock SCF Theory"

1963

F.W. Birss, S. Fraga; J. Chem. Phys., 38, 2552 (1963)
"SCF Theory. I. General Treatment"

R. McWeeny; Quant. Chem. Group for Research in Atomic Molecular and Solid-State Theory, Uppsala University, Preprint No. 98 (1963)
"SCF Theory of Open Shell Systems"

1964

F.W. Birss, S. Fraga; J. Chem. Phys., 40, 3212 (1964)
"SCF Theory. IV. LCAO Approximation for Excited States"

F.W. Birss, W.G. Laidlaw; Theoret. Chim. Acta, 2, 186 (1964)
"On the Empirical Validity of Koopman's Theorem"

S. Fraga, F.W. Birss; J. Chem. Phys., 40, 3203 (1964)
"SCF Theory. II. The LCAO Approximation"

S. Fraga, F.W. Birss; J. Chem. Phys., 40, 3207 (1964)
"SCF Theory. III. General Treatment for Excited States"

S. Fraga; Theoret. Chim. Acta, 2, 403 (1964)
"Non-Relativistic SCF Theory. I."

S. Fraga; Theoret. Chim. Acta, 2, 406 (1964)
"Non-Relativistic SCF Theory. II."

S. Fraga; Theoret. Chim. Acta, 2, 411 (1964)
"Non-Relativistic SCF Theory. III."

W.G. Laidlaw, F.W. Birss; Theoret. Chim. Acta, 2, 181 (1964)
"Koopman's Theorem and Virtual Orbital Energies in the General SCF Theory"

R.E.D. McClung, S. Fraga; Theoret. Chim. Acta, 2, 416 (1964)
"Non-Relativistic SCF Theory. IV."

1965

M. Cohen, P.S. Kelly; Can. J. Phys., 43, 1867 (1965)
"Hartree-Fock Wave Functions for Excited States. The $2\,^1S$ State of He"

S. Huzinaga; Memoirs Faculty of Science. Kyusyu University, Ser. B, 3, 73 (1965)
"SCF Method in the Paired-Electron Approximation"

1966

G. Verhaegen, W.G. Richards; J. Chem. Phys., 45, 1828 (1966)
"Valence Levels of Beryllium Oxide"

1967

Ch. Froese-Fischer; Can. J. Phys., 45, 7 (1967)
"The orthogonality assumption in the Hartree-Fock approximation"

H.W. Kroto, D.P. Santry; J. Chem. Phys., 47, 792 (1967)
"CNDO MO Theory of Molecular Spectra. I. The Virtual-Orbital Approximation to Excited States"

H.W. Kroto, D.P. Santry; J. Chem. Phys., 47, 2736 (1967)
"Semiempirical MO Theory of Molecular Spectra. II. Approximate Open-Shell Theory"

1968

D.H. Sleeman; Theoret. Chim. Acta., 11, 135 (1968)
"The Determination of SCF LCAO Solutions for Open Shell Configurations"

M. Yaris, A. Moscowitz, R.S. Berry; J. Chem. Phys., 49, 3150 (1968)
"Low-Lying Excited States of Mono-olefins"

1969

W.A. Goddard III, T.H. Dunning, Jr., W.J. Hunt; Chem. Phys. Lett., 4, 231 (1969)
"The proper treatment of off-diagonal Lagrange Multipliers and Coupling Operators in Self-Consistent Field Equations"

W.J. Hunt. T.H. Dunning, Jr., W.A. Goddard III; Chem. Phys. Lett., 3, 606 (1969)
"The orthogonality constrained Basis Set Expansion Method for treating off-diagonal Lagrange Multipliers in calculations of Wave Functions"

W.H. Hunt, W.A. Goddard; Chem. Phys. Lett., 3, 414 (1969)
"Excited States of H_2O using Improved Virtual Orbitals"

S. Huzinaga; J. Chem. Phys., 51, 3971 (1969)
"Coupling Operator Method in the SCF Theory"

G.W. King, D.P. Santry, C.H. Warren; J. Chem. Phys., 50, 4565 (1969)
"CNDO MO Theory of Molecular Spectra III Calculations for the Fluorosulfate Radical"

G.H. Kirby, K. Miller; Chem. Phys. Lett., 3, 643 (1969)
"The CNDO Geometry of C_2H_4 in the first singlet excited State"

1970

H. Basch, V. McKoy; J. Chem. Phys., 53, 1628 (1970)
"Interpretation of Open-Shell SCF Calculations on the T and V States of Ethylene"

P.J. Bertoncini, G. Das, A.C. Wahl; J. Chem. Phys., 52, 5112 (1970)
"Theoretical Study of the $^1\Sigma^+$, $^3\Sigma^+$, $^3\Pi$, $^1\Pi$ States of NaLi and the $^2\Sigma^+$ State of $NaLi^+$"

Th.M. Brown, W.F. Myszkowski; J. Chem. Phys., 52, 4918 (1970)
"Rapid convergence in open shell SCF π electron calculations"

J. Cizek, J. Paldus; J. Chem. Phys., 53, 821 (1970)
"Stability Conditions for the Solutions of the Hartree-Fock Equations for Atomic and Molecular Systems. III. Rules for the Singlet Stability of Hartree-Fock Solutions of π-Electronic Systems"

J.P. Dahl, H. Johansen, D.R. Truax, T. Ziegler; Chem. Phys. Lett., 6, 64 (1970)
"On the Derivation of Necessary Conditions on Hartree-Fock Orbitals"

I.H. Hillier, V.R. Saunders; Int. J. Quant. Chem., 4, 503 (1970)
"A New SCF Procedure and its Applications to Ab Initio Calculations of the States of the Fluorosulfate Radical"

W.J. Hunt, W.A. Goddard III, T.H. Dunning, Jr.; Chem. Phys. Lett., 6, 147 (1970)
"Incorporation of quadratic convergence into open-shell SCF equations"

R. Kari, B.T. Sutcliffe; Chem. Phys. Lett., 7, 149 (1970)
"Direct minimization of the energy functional in some open-shell LCAO calculations on atom"

J. Paldus, J. Cizek; J. Chem. Phys., 52, 2919 (1970)
"Stability Conditions for the Solutions of the Hartree-Fock Equations for Atomic and Molecular Systems. II. Simple Open-Shell Case"

G.A. Segal; J. Chem. Phys., 52, 3530 (1970)
"Alternative Technique for the Calculation of Single-Determinant Open-Shell SCF Functions which are Eigenvalues of S^2"

G.A. Segal; J. Chem. Phys., 53, 360 (1970)
"Calculation of Wavefunctions for the Excited States of Polyatomic Molecules"

R. Zakradnik, P. Carsky; J. Phys. Chem., 74, 1235 (1970)
"Conjugated radicals. I. Introductory remarks and method of calculation"

1971

P. Carsky, R. Zahradnik; Collect. Czech. Chem. Comm., 71, 961 (1971)
"Conjugated radicals. VIII. Comparison of the Open Shell SCF results obtained by the method of Longuet-Higgins and Pople, and by the method of Roothaan"

C.A. Coulson; Mol. Phys., 20, 687 (1971)
"Brillouin's theorem and the Hellman-Feynman Theorem for Hartree-Fock wave functions"

F.O. Ellison, F.M. Matheu; Chem. Phys. Lett., 10, 322 (1971)
"Generalization of Dewar's Half-Electron Method for Calculating Energies of Open-Shell Electronic States"

G.H. Kirby, K. Miller; J. Mol. Struct., 8, 373 (1971)
"Computation of Molecular Geometries for Singlet Excited States"

G. Roberts, K.D. Warren; Theoret. Chim. Acta, 22, 184 (1971)
"SCF-π-electron calculations using orthogonalized AO"

J.B. Rose, V. McKoy; J. Chem. Phys., 55, 5435 (1971)
"Applicability of SCF theory to some open-shell states of CO, N_2 and O_2"

J.W. Richardson, Th.F. Soules, D.M. Kaught, R.R. Powell; Phys. Rev., B 4, 1721 (1971)
"Open-shell SCF MO Theory for transition metal clusters"

T.E.H. Walker; Chem. Phys. Lett., 9, 174 (1971)
"Sources of error in Open Shell HF calculations"

M.A. Whitehead, D.H. Lo; J. Chem. Soc., A, 463 (1971)
"Semiempirical MO calculations on closed and open-shell hydrocarbons"

R. Zahradnik, S. Huenig, D. Schentzow, P. Carsky; J. Phys. Chem., 75, 335 (1971)
"Conjugated radicals. IV. Experimental study and the LCI SCF open shell calculations on the electronic spectra and the redox equilibria of the nitrogen containing violenes"

1972

P. Carsky, R. Zahradnik; Theoret. Chim. Acta, 26, 171 (1972)
"A Remark on the Comparison between the Roothaan Open-Shell and Half-Electron Method"

J.L. Dodds, R. McWeeny; Chem. Phys. Lett., 13, 9 (1972)
"Orbital Energies and Koopman's Theorem in Open-Shell Hartree-Fock Theory"

I. Fischer-Hjalmars, J. Kowalewski; Theoret. Chim. Acta, 27, 197 (1972)
"Simplified Non-Empirical Excited State Calculations": I. The Rydberg ns and $np\sigma$ Series of Ethylene"

T. Fueno; Kagaku (Kyoto), 27, 889 (1972)
"Electronic Structure in open shell systems"

P.J. Hay, W.J. Hunt, W.A. Goddard; Chem. Phys. Lett., 13, 30 (1972)
"Generalized Valence Bond Wavefunctions for the Low Lying States of Methylene"

P Jørgensen; J. Chem. Phys., 57, 4884 (1972)
"Electronic Excitations of Open Shell systems in the Grand Canonical and Canonical Time-Dependent Hartree-Fock Models. Applications on Hydrocarbon Radical Ions"

J. Langlet; Theoret. Chim. Acta, 27, 223 (1972)
"PCILO Method for Excited States. I. Construction of the Zeroth Order Wave-Function for Planar Conjugated Systems"

R. Manne; Mol. Phys., 24, 935 (1972)
"Brillouin's Theorem in Roothaan's open-shell SCF method"

T. Morikawa; Chem. Phys. Lett., 17, 297 (1972)
"Method of the orthonormality-constrained variation"

K. Morokuma, S. Iwata; Chem. Phys. Lett., 16, 192 (1972)
"Extended Hartree-Fock Theory for Excited States"

D. Peters; J. Chem. Phys., 57, 4351 (1972)
"Simple procedure for Open Shell SCF MO Computations"

D.H. Phillips, J.C. Schug; J. Chem. Phys., 57, 3498 (1972)
"Projected States of Open Shell Molecules: The Pi-Electron States of the Cyclopentadienyl Cation"

T.E.H. Walker; Theoret. Chim. Acta, 25, 1 (1972)
"Open-Shell HF calculations"

R. Zahradnik, P. Carsky; Theoret. Chim. Acta, 27, 121 (1972)
"Open Shell CNDO Treatments on Small and Aliphatic Radicals. Electronic Spectra and Some Ground State Properties"

1973

R. Albat, N. Gruen; Chem. Phys. Lett., 18, 572 (1973)
"Examples of known SCF procedures which do not satisfy all necessary conditions for the energy to be stationary"

R. Albat, N. Gruen; J. Phys. (B) Atom. Molec. Phys., 6, 601 (1973)
"A derivation of general Open-Shell SCF Equations for atoms and molecules with a Green Function Method"

M. Benard, A. Julg; Int. J. Quant. Chem., 7, 945 (1973)
"Determination des Orbitales Moléculaires par Minimisation Directe de l'Energie de L'Etat Fondamental et des Etats Excités des Molécules"

R.R. Birge; J. Am. Chem. Soc., 95, 8241 (1973)
"Photochemical and spectroscopic applications of approximate MO theory. I. Averaged field approximate open-shell theory"

C. Bottcher; J. Phys. (B) Atom. Molec. Phys., 6, 2368 (1973)
"Model potential Calculations on Open-Shell Molecules"

P. Carsky, M. Machacek, R. Zahradnik; Collect. Czech. Chem. Commun., 38, 3067 (1973)
"Open Shell CNDO Treatments on Small Inorganic Radicals"

H.M. Chang; Diss. Abstr. Int. B, 1974, 34, 4306 (1973)
"Study of open-shell systems by the CNDO/S method"

H.M. Chang, H.H. Jaffe; Chem. Phys. Lett., 23, 146 (1973)
"Uses of the CNDO Method in Spectroscopy. Doublet States"

J.C. Cobb, A. Hinchlife; J. Mol. Struct., 15, 1 (1973)
"Open Shell SCF Calculations for Non-Orthogonal Basis Functions"

E.R. Davidson; Chem. Phys. Lett., 21, 565 (1973)
"Spin-Restricted Open-Shell SCF Theory"

D.W. Davies, G. Del Conde; Rev. Mex. Fis., 22, 343 (1973)
"Open Shell HF methods for molecules. Hydrogen molecule anion"

T.D. Davis, R.E. Christoffersen, G.M. Maggiora; Chem. Phys. Lett., 21, 576 (1973)
"Ab Initio Calculations on Large Molecules Using Molecular Fragments, Initial Studies on Open Shell Systems"

I. Fischer-Hjalmars, J. Kowalewski; Theoret. Chim. Acta, 29, 345 (1973)
"Simplified Non-Empirical Excited State Calculations: II. Interpretations of the Electronic Transitions in the Vacuum UV Spectrum of Ethylene"

H. Fukutome; Prog. Theoret. Phys., 50, 1433 (1973)
"The unrestricted Hartree-Fock Theory of Chemical Reactions III"

K. Hirao, H. Nakatsuji; J. Chem. Phys., 59, 1457 (1973)
"General SCF Operator satisfying correct variational Condition"

R. Kari, B.T. Sutcliffe; Int. J. Quant. Chem., 7, 459 (1973)
"Direct Minimization of the Energy Functional in LCAO-MO Calculations"

R. Macaulay, D. Croutier; Chem. Phys. Lett., 18, 501 (1973)
"An Open-Shell SCF Program for use in Ab Initio Gaussian-Type Calculations"

O. Sinanoglu; J. Mol. Struct., 19, 81 (1973)
"Prediction of Molecular Excited State Properties, Potential Energy Curves, and the Non-Closed Shell Many-Electron Theory"

N.F. Stepanov, A.A. Ustenko, A.I. Demet'ev; Vestn. Mosk. Univ. Khim., 14, 102 (1973)
"Ionization potentials of open-shell molecules"

D.G. Truhlar; Int. J. Quant. Chem., 7, 1175 (1973)
"Ab Initio Hartree-Fock Calculations for Electronic Wave Functions for the $c^3\Pi_\mu$ State of H_2"

P.S.C. Wang, M.L. Benston, D.P. Chong; J. Chem. Phys., 59, 1721 (1973)
"Constrained Variation Method for Excited-State Energies of Atoms and Molecules"

1974

N.C. Baird, R.F. Barr; Theoret. Chim. Acta, 36, 125 (1974)
"Comparison of different methods in determining energies and geometries of open-shell systems"

P.M. Blustin, J.W. Linnett; J. Chem. Soc. Far. Trans II, 70, 327 (1974)
"Applications of a simple molecular wavefunction. 5. Floating spherical Gaussian Orbital open-shell calculations. Introduction"

P.M. Blustin, J.W. Linnett; J. Chem. Soc. Far. Trans II, 70, 826 (1974)
"Applications of a simple molecular function. 6. FSGO, open-shell calculations on first-row diatomic molecular systems"

B.L. Burrows, A.T. Amos; Theoret. Chim. Acta, 36, 1 (1974)
"Perturbation Theory for Excited States of Molecules. 5. Polarizabilities of Singlet and Triplet Excited States of Conjugated Molecules"

R. Caballol, R. Gallifa, J.M. Riera, R. Carbo; Int. J. Quant. Chem., 8, 373 (1974)
"Generalized Open Shell SCF Theory"

A.I. Dement'ev, N.F. Stepanov, S.S. Yarovoi; Int. J. Quant. Chem., 8, 107 (1974)
"Convergence Problems in the Solution of SCF Equations"

B.J. Duke, J.E. Eiliers, B. O'Leary; J. Chem. Soc. Far. Trans II, 70, 386 (1974)
"Simulated Ab Initio MO Technique. III Open Shell Radicals in the Spin Unrestricted Formalism"

E. Gey, Ch. Jung, J. Sauer; Collect. Czech. Chem. Commun., 39, 1235 (1974)
"Restricted HF calculations of open-shell systems by means of semi-empirical MO LCAO SCF procedures"

T.L. Gilbert; J. Chem. Phys., 60, 1789 (1974)
"Selection of SCF Orbitals"

M.F. Guest, V.R. Saunders; Mol. Phys., 28, 819 (1974)
"On Methods for Converging Open-Shell Hartree-Fock Wave-Functions"

K. Hirao; J. Chem. Phys., 60, 3215 (1974)
"On the Coupling Operator Method"

K. Hirao; J. Chem. Phys., 60, 3247 (1974)
"How to Resolve the Orbital Ambiguity to Obtain the Orbital Set which is Stable to an Excitation"

C.F. Jackels, E.R. Davidson; Int. J. Quant. Chem., 8, 707 (1974)
"Equivalence-Restricted Open-Shell SCF Theory"

J. Kuhn, P. Carsky, R. Zahradnik; Theoret. Chim. Acta, 33, 263 (1974)
"SCF-CI MO Treatment of Radicals Having Degenerate Ground States"

J. Langlet, J.P. Malrieu; Theoret. Chim. Acta, 33, 307 (1974)
"Geometries of Excited States of Small Polyenes"

R. McWeeny; Mol. Phys., 28, 1273 (1974)
"SCF Theory for Excited-States. I. Optimal Orbitals for the States of a Configuration"

N. Moiseyev, J. Katriel; Chem. Phys. Lett., 29, 69 (1974)
"The Continuity Dilemma and Hartree-Fock Inestabilities"

S. Narita, K. Sazi, Y.J. I'Haya; Chem. Phys. Lett., 29, 232 (1974)
"A "Pseudo-Potential" Study on Roothaan's Open-Shell System"

B.T. Sutcliffe; Theoret. Chim. Acta, 33, 201 (1974)
"The Convergence Properties of Direct Methods of Energy Minimization with Respect to Linear Coefficients in the LCAO-MO-SCF Approach"

K.F. Treed; Chem. Phys., 4, 80 (1974)
"Open-Shell Generalized Perturbation Theory"

M.H. Wood; Chem. Phys. Lett., 28, 477 (1974)
"On the Calculations of the Excited States of Small Molecules"

D.R. Yarkony, H.F. Schaefer III; J. Am. Chem. Soc., 96, 3754 (1974)
"Triplet Electronic Ground State of Trimethylenemethane"

1975

R.D. Amos, G. Doggett; Mol. Phys., 29, 1117 (1975)
"Spin-Coupled Wavefunctions I. The three-electron Doublet States"

R.D. Amos; Mol. Phys., 29, 1125 (1975)
"Spin-Coupled Wavefunctions II. The four-electron Singlet and Triplet States"

R. Ahlrichs; Chem. Phys. Lett., 34, 570 (1975)
"Hartree-Fock Theory for Negative Ions"

M. Ahmed, L. Lipsky; Phys. Rev., A12, 1176 (1975)
"Triply Excited States of Three-Electron Atomic Systems"

N.L. Arthur, K.F. Donchi, J.D. McDonell; J. Chem. Phys., 62, 1585 (1975)
"BEBO Calculations. III. A new Triplet Repulsion Energy Term"

P.M. Blustin, J.W. Linnett; J. Chem. Soc. Far. Trans. II, 71, 1058 (1975)
"Applications of a simple molecular wave function. 7. FSGO open shell calculations on first-row polyatomic hydrides and hydride ions"

R. Carbo, R. Gallifa, J.M. Riera; Chem. Phys. Lett., 30, 43 (1975)
"Some Remarks about a Generalized SCF Coupling Operator Open Shell Theory"

P. Carsky, J. Kahn, R. Zahradnik; J. Mol. Spectroscopy, 55, 120 (1975)
"Semiempirical All-Valence-Electron MO Calculations on the Electronic Spectra of Linear Radicals with degenerate Ground States"

R. Constanciel, O. Chalvet, J.C. Rayez; Theoret. Chim. Acta, 37, 305 (1975)
"Comparative Study of the pK of Anidine, Thionine and Phenazine Molecules in their First Excited Singlet and Triplet States"

D. Demoulin; Chem. Phys., 11, 329 (1975)
"The Shapes of Some Excited States of C_2H_2"

W.B. England, N.H. Sabelli, A.C. Wahl; J. Chem. Phys., 63, 4596 (1975)
"A Theoretical Study of Li_2H. I. Basis set and Computational Survey of Excited States and Possible Reaction Paths"

M.F. Guest, V.R. Saunders; Mol. Phys., 29, 873 (1975)
"The Calculation of Valence Shell Ionization Potentials by the ΔSCF Method"

W.G. Herkstroeter; J. Am. Chem. Soc., 97, 4161 (1975)
"The Triplet Energies of Azulene, β-Carotene, and Fenocene"

Y. Horino, H. Tatewaki; Int. J. Quant. Chem., 9, 287 (1975)
"The Electronic Structure of the Excited States for B, C^+, and C Arising from the Configuration $1s^22s2p^n$. The Effect of Removal of the "Equivalence" Restriction"

G.A. Hart, P.L. Goodfriend; Mol. Phys., 29, 1109 (1975)
"Model Pseudopotential open shell LCAO-MO-SCF calculations on alkali metal trimers"

T. Kagawa; Phys. Rev., A12, 2245 (1975)
"Relativistic HFR Theory for Open-Shell Atoms"

G.H. Kirby; J. Mol. Struct., 26, 77 (1975)
"Optimization of Excited State Geometries of Fulminic Acid and Isocyanic Acid"

D.A. Luippold; Chem. Phys. Lett., 35, 131 (1975)
"INDO Theoretical Studies of the Lowest Triplet and Singlet States of Stilbene"

O. Matsouka; Mol. Phys., 30, 1293 (1975)
"Equivalence of Saunder's Level-Shifting Method for Hartree-Fock Equations and Expansion Method for Modified Hartree-Fock Equations"

O. Matsuoka, H. Ito; Theoret. Chim. Acta, 39, 111 (1975)
"Random-Phase-Approximation Calculations on Triplet Spectra of Conjugated Molecules"

R. McWeeny; Chem. Phys. Lett., 35, 13 (1975)
"Effective Hamiltonians and Orbital Optimization"

M.M. Mestechkin, G.E. Whyman; Int. J. Quant. Chem., 9, 761 (1975)
"Density Matrix in the Open Shell Theory"

D. Mukherjee; Int. J. Quant. Chem., 9, 943 (1975)
"Orbital Optimization in Single Open-Shell Configurations: A Sequential Unconstrained Minimization Technique"

D. Mukherjee, R.K. Moitra, A. Mukhopadhyay; Pramana, 4, 247 (1975)
"A non-perturbative Open-Shell Theory for Atomic and Molecular Systems: Application to Transbutadiene"

D. Mukherjee, R.K. Moitra, A. Mukhopadhyay; Mol. Phys., 30, 1861 (1975)
"Correlation Problem in Open-Shell Atoms and Molecules. A non-perturbative Linked Cluster Formulation"

H.P. Roy, A. Gupta, P.K. Mukherjee; Int. J. Quant. Chem., 9, 75 (1975)
"Frequency-Dependent Polarizability of Open-Shell Atomic Systems"

J. Sauer, C. Jung; Theoret. Chim. Acta, 40, 129 (1975)
"Consequences of Koopman's Theorem in the Restricted HF Methods for Open Shell Systems"

R.F. Steward; Mol. Phys., 30, 1283 (1975)
"A numerical Study of Coupled Hartree-Fock Theory for Open-Shell Systems"

O. Sinanoglu, D.R. Herrick; Chem. Phys. Lett., 31, 373 (1975)
"New Evidence for a Possible Hidden Symmetry in Doubly Excited He"

R.F. Steward; Mol. Phys., 30, 1283 (1975)
"A numerical Study of Coupled Hartree-Fock Theory for Open-Shell Systems"

P. Winkler, R.N. Porter; J. Chem. Phys., 62, 257 (1975)
"Natural Orbitals of Several Excited States of the He atom"

K. Yamaguchi; Chem. Phys. Lett., 33, 330 (1975)
"The Electronic Structures of Biradicals in the Unrestricted Hartree-Fock Approximation"

D.L. Yeager and V. McKoy; J. Chem. Phys., 63, 4861 (1975)
"An Equations of Motion Approach for Open Shell Systems"

V.M. Zelichenko, A.F. Terpugova, E.I. Cheglokov; Kvant. Khimiya, 3, (1975)
"Excited states of molecules with open shells in a frozen core model"

76

D.R. Beck, C.A. Nicolaides; Int. J. Quant. Chem., 10, 119 (1976)
"Theory and Calculation of Excited-State Wave Functions and Properties"

P. Bischof; J. Am. Chem. Soc., 98, 6844 (1976)
"Unrestricted open-shell calculations by MINDO/3. Geometries and electronic structure of radicals"

J.W. Caldwell, M.S. Gordon; Chem. Phys. Lett., 43, 493 (1976)
"SCF Calculations on Excited States"

M.P.S. Collins, B.J. Duke, J.E. Eilers, B. O'Leary; Int. J. Quant Chem., 10, 629 (1976)
"The Simulated Ab Initio Molecular Orbital Technique. VI. Open-Shell Radicals in the Spin Restricted Formalism"

E.R. Davidson, L.Z. Stenkamp; Int. J. Quant. Chem., 10, 21 (1976)
"SCF Methods for Excited States"

K. Faegri, R. Manne; Mol. Phys., 31, 1037 (1976)
"A new Procedure for Roothaan's symmetry-restricted Open Shell SCF Method"

F. Flouquet; Chem. Phys., 13, 257 (1976)
"Ab Initio Study of the Potential Energy Surface of the Lowest A_1 Symmetry excited State of H_2S"

M.S. Gordon; Chem. Phys. Lett., 37, 593 (1976)
"Excited States and Photochemistry of Saturated Molecules. Methylene Fluoride"

J.E. Grabenstetter, F. Grein; Mol. Phys., 31, 1469 (1976)
"Improved Convergence of Open-Shell SCF Calculations Level-Shifting in the Double-Matrix Formalism"

J. Hendekovic; Theoret. Chim. Acta, 42, 193 (1976)
"General Procedure for Solving the Open-Shell SCF Secular Equations"

H. Hsu, E.R. Davidson, R.M. Pitzer; J. Chem. Phys., 65, 609 (1976)
"An SCF Method for Hole States"

H.H. Huang, J.W. Linnett; J. Chem. Soc. Chem. Commun., 76, 135 (1976)
"FSGO open-shell calculations on linear triatomic and tetratomic hydrogen complexes"

Y. Ishikawa; Chem. Phys. Lett., 37, 597 (1976)
"Ab Initio SCF Calculations by a Generalized Coupling Operator Method"

H.H. Jaffe, J. Singerman; J. Phys. Chem., 80, 1928 (1976)
"Orbital energies in open shell systems"

J.M. Leclerq, C. Mijoule, P. Yvan; J. Chem. Phys., 64, 1464 (1976)
"Theoretical Investigations of Excited States of Glyoxal and Biacetyl"

S. Nagare, T. Fueno; Theoret. Chim. Acta, 41, 9 (1976)
"Reaction paths of some open-shell reactive intermediates. An intermolecular perturbation approach"

A.H. Pakiari, J.W. Linnett; J. Chem. Soc. Far. Trans. II, 72, 641 (1976)
"Applications of a simple molecular wave function. Part 11. Extension of FSGO method to open-shell treatments"

A.H. Pakiari, J.W. Linnett; J. Chem. Soc. Far. Trans. II, 72, 1281 (1976)
"Applications of a simple molecular wave function. Part 12. Open-shell floating spherical Gaussian orbital calculations for some atoms and ions"

A.H. Pakiari, J.W. Linnett; J. Chem. Soc. Far. Trans. II, 72, 1288 (1976)
"Applications of a simple molecular wave function. Part 13. Open-shell calculations for hydrogen-bridge structures"

K. Yamaguchi, T. Fueno; Chem. Phys. Lett., 38, 47 (1976)
"The unstability conditions of the Restricted Hartree-Fock (RHF) Solutions in the Doublet State"

K. Yamaguchi, T. Fueno; Chem. Phys. Lett., 38, 52 (1976)
"Spin-Symmetry Forbidden Properties of Free-Radical Cydoadditions"

D.R. Yarkoni; Diss. Abstr. Int. B., 37, 258 (1976)
"Advances in the solution of the electronic Schroedinger equation for moderately symmetric open shell molecules"

Bibliographical Survey on Multiconfigurational SCF Theory

1965

T.L. Gilbert; J. Chem. Phys., 43, 5248 (1965)
"Optimum-Multiconfiguration Self-Consistent-Field Equations"

1966

G. Das, A.C. Wahl; J. Chem. Phys., 44, 87 (1966)
"Extended HF Wave functions: Optimized Valence Configurations for H_2 and Li_2, Optimized Double Configurations for F_2"

1967

W.H. Adams; Phys. Rev., 156, 109 (1967)
"Orbital SCF Theory I. General Theory for MC Wave Functions"

E. Clementi; J. Chem. Phys., 46, 3842 (1967)
"Study of the Electronic Structure of Molecules. I. Molecular Wavefunctions and their Analysis"

G. Das; J. Chem. Phys., 46, 1568 (1967)
"Extended HF Ground-State Wavefunctions for the Lithium Molecule"

G. Das, A.C. Wahl; J. Chem. Phys., 47, 2934 (1967)
"Extended HF Wavefunctions: General Theory of Optimized-Valence Configurations and Its Application to Diatomic Molecules"

F.E. Harris; J. Chem. Phys., 46, 2769 (1967)
"Open-Shell Orthogonal Molecular Orbital Theory"

J. Hinze, C.C.J. Roothaan; Prog. Theoret. Phys. Suppl., 40 (1967)
"Multi-Configuration SCF Theory"

A. Veillard, E. Clementi; Theoret. Chim. Acta, 7, 133 (1967)
"Complete MC SCF Theory"

W.R. Wessel; J. Chem. Phys., 47, 3253 (1967)
"Iterative Quadratically Convergent Algorithm for Solving the General HFR Equations"

1968

M.L. Benston, D.P. Chong; Mol. Phys., 14, 449 (1968)
"MC SCF Theory with non-Orthogonal orbitals"

B. Levy; J. Chem. Phys., 48, 1994 (1968)
"Best choice for the Coupling Operators in the Open-Shell and MC SCF Methods"

B. Levy, G. Berthier; Int. J. Quant. Chem., 2, 307 (1968)
"Generalized Brillouin Theorems for MC SCF Theories"

R. McWeeny; Simposia Far. Soc., 2, 7 (1968)
"MC SCF Calculations"

1969

P.S. Bagus, N. Bessis, C.M. Moser; Phys. Rev., 179, 39 (1969)
"MC HF Calculations. II. Calculation of the Lowest ^3P, ^1D and ^1S States of the Carbon Atom"

P.S. Bagus, C.M. Moser; J. Phys. B (Atom. Molec. Phys.), 2, 1214 (1969) Ser. 2
"MC HF Calculations. III. Calculations of the ^3P, ^1D and ^1S States Arising from the $1s^2 2s^2 2p^2$ Configuration for Z = 7.0 to Z = 30.0"

Ch. Froese-Fischer; Comp. Phys. Commun., 1, 151 (1969)
"A MC HF Program"

S. Huzinaga; Prog. Theoret. Phys., 41, 307 (1969)
"A General SCF Formalism"

N. Sabelli, J. Hinze; J. Chem. Phys., 50, 684 (1969)
"Atomic MC SCF Wavefunctions"

W.T. Zemke, P.G. Lykos, A.C. Wahl; J. Chem. Phys., 51, 5635 (1969)
"Double Configuration SCF STudy of the $^1\Pi_\mu$, $^3\Pi_\mu$, $^1\Pi_g$, and $^3\Pi_g$ States of H_2"

1970

P.J. Bertoncini, G. Das, A.C. Wahl; J. Chem. Phys., 52, 5112 (1970)
"Theoretical Study of the $^1\Sigma^+$ $^3\Sigma^+$, $^3\Pi$, $^1\Pi$ States of NaLi and the $^2\Sigma^+$ State of $NaLi^+$"

S. Fraga, Y.G. Smeyers; Anales de Fisica, 66, 259 (1970)
"Simultaneous MC-SCF Theory"

Ch. Froese-Fischer; J. Phys. B. 3, 779 (1970)
"Comparison of MC HF and CI results for the ^1D series of Mg"

D. Fu-Tai Tuan; J. Chem. Phys., 52, 5247 (1970)
"Error of Expectation Values from the MC-SCF Theory"

Yu. A. Kruglyak, V.A. Kuprievich, E.V. Mozdor; Str. Mol. Krantovaya Khim., 121 (1970)
"Calculation of wavefunctions of molecules in a multiconfigurational approximation"

B. Levy; Int. J. Quant. Chem., 4, 297 (1970)
"Molecular MC-SCF Calculations"

N.G. Mukherjee, R. McWeeny; Int. J. Quant. Chem., 4, 97 (1970)
"MC SCF Calculations on LiH"

A.C. Wahl, G. Das; Advan. Quantum Chem., 5, 261 (1970)
"Optimized valence configurations: A reasonable application of the MCSCF technique to the quantitative description of chemical bonding"

1971

H. Basch; J. Chem. Phys., 55, 1700 (1971)
"Dimerization of Methylenes by Their Least Motion, Coplanar Approach: A MC-SCF Study"

J.McI. Calvert, W.D. Davidson; J. Phys. B., 4, 314 (1971)
"Multiconfiguration frozen core approximation. I. Eigenstates of the two-electron atom"

F. Grein, T.C. Chang; Chem. Phys. Lett., 12, 44 (1971)
"MC wavefunctions obtained by application of the generalized Brillouin theorem"

A. Kancerevicius; Liet. Fiz. Rinkins, 11, 401 (1971)
"MC approximation for the ns^2 shells of atoms"

V. Kaveckis, V. Kaminskas; Liet. Fiz. Rinkins, 11, 197 (1971)
"Extended method of calculation and multiconfiguration approximation for He-like atoms"

N.G. Mukherjee; Indian J. Pure Appl. Phys., 71, 70 (1971)
"Convergence of MCSCF optimization methods"

H.F. Schaefer III, C.F. Bender; J. Chem. Phys., 55, 1720 (1971)
"MC Wavefunctions for the H_2O Molecule"

R.M. Stevens; J. Chem. Phys., 55, 1725 (1971)
"Accurate SCF Calculation for NH_3 and its Inversion Motion"

1972

Ch.F. Bender, Th. M. Dunning, M.F. Schaefer, N.A. Goddard, W.J. Hunt; Chem. Phys. Letts., 15, 171 (1972)
"MC wavefunctions for the lowest (π-π*) excited states of ethylene"

T.C. Chang, F. Grein; J. Chem. Phys., 57, 5270 (1972)
"Two-By-Two Rotation Method Applied to Selected MC Wavefunctions"

G. Das, A.C. Wahl; J. Chem. Phys., 56, 1769 (1972)
"New Techniques for the Computation of MC-SCF Wavefunctions"

G. Das, A.C. Wahl; J. Chem. Phys., 56, 3532 (1972)
"Theoretical Study of the F_2 Molecule Using the Method of Optimized Valence Configurations"

P. Dejardin, E. Kochanski, A. Veillard; Chem. Phys. Lett., 15, 248 (1972)
"MC SCF calculation for the NH_3 molecule"

Ch. Froese-Fischer; Comp. Phys. Commun., 4, 107 (1972)
"A MC HF Program with Improved Stability"

T.L. Gilbert; Phys. Rev., A6, 580 (1972)
"MC SCF Theory for Localized Orbitals. I. The Orbital Equations"

A. Kancerevicius; Liet. Fiz. Rinkinys, 12, 357 (1972)
"MC approximation for atoms with outer $2p^n$ shells"

V.A. Kuprievich, O.V. Shramko; Int. J. Quant. Chem., 6, 327 (1972)
"MC SCF theory. Method of one-electron Hamiltonian"

W. Kutzelnigg, V. Staemmler, M. Gelus; Chem. Phys. Lett., 13, 496 (1972)
"Potential Curve of the Lowest Triplet State of Li_2"

N.G. Mukherjee; Indian J. Pure Appl. Phys., 10, 594 (1972)
"Optimization of MC SCF orbitals"

H. Schlosser; J. Chem. Phys., 57, 4332 (1972)
"Simplified MC SCF Theory for Localized Orbitals. I. Fixed Orbitals"

H. Schlosser; J. Chem. Phys., 57, 4342 (1972)
"Simplified MC SCF Theory for Localized Orbitals. II. Fluctuating Orbitals"

Z. Sibincic; Phys. Rev., A5, 1150 (1972)
"MC SCF calculations for several states of B"

1973

P.S. Bagus, J. Bauche; Phys. Rev., A8, 734 (1973)
"Evaluation of the Orbital-Dependent Hyperfine Constants of the 2p-Series Atoms from MC HF Wave Functions"

H. Basch; Chem. Phys. Lett., 19, 323 (1973)
"Open-Shell MC SCF Results for the Lowest Energy $^1(\pi,\pi^*)$ State of Planar Ethylene"

G. Das; J. Chem. Phys., 58, 5104 (1973)
"MC SCF Theory for Excited States"

P. Dejardin, E. Kochanski, A. Veillard, B. Roos, P. Siegbahn; J. Chem. Phys., 59, 5546 (1973)
"MC-SCF and CI Calculations for the NH_3 Molecule"

Ch. Froese-Fischer; J. Phys. (B) Atom. Molec. Phys., 6, 1933 (1973)
"Brillouin's Theorem for excited $(n\ell)^q\,(n'\ell)^{q'}$ Configurations"

Ch. Froese-Fischer; Can. J. Phys., 51, 1238 (1973)
"MC HF Correlation Study of $1s2s\,^1S$"

Ch. Froese-Fischer; J. Comput. Phys., 13, 502 (1973)
"Solution of Schroedinger equation for two-electron systems by an MCHF procedure"

F. Grein, T.C. Chang; J. Phys. (B) Atom. Molec. Phys., 6, L237 (1973)
"Single Excitations in MC Wavefunctions: Ground States of the Atoms Be to F"

A. Golebiewski, E. Novak-Broclawik; Mol. Phys., 26, 989 (1973)
"On the MC SCF Theory of Closed-Shell Systems. I. Physical Model and Generalized Brillouin Theorem"

H. Hinze; J. Chem. Phys., 59, 6424 (1973)
"On Convergence Guarantees for the MC SCF Theory"

R.P. Hosteny, A.R. Huids, A.C. Wahl, M. Krauss; Chem. Phys. Lett., 23, 9 (1973)
"MC SCF Calculations of the lowest Triplet State of H_2O"

B. Huron, J.P. Malrieu, P. Rancurel; J. Chem. Phys., 58, 5745 (1973)
"Iterative Perturbation Calculations of Ground and Excited State Energies from MC zeroth-order Wavefunctions"

K. Ishida, H. Nakatsuji; Chem. Phys. Lett., 19, 268 (1973)
"MC SCF Wavefunctions for the Fermi-Contact Hyperfine Structure of Li Atom"

K. Jug; Theoret. Chim. Acta, 30, 231 (1973)
"Semiempirical extended HF theory"

A. Kairo, M. Krauss, A.C. Wahl; Int. J. Quant. Chem. Symp., 7, 143 (1973)
"Recent Applications of the MC SCF Method to Polarizabilities, Excited States, Van der Waals Forces, and Triatomic Surfaces"

A. Kancerevicius, V.A. Bolotin; Liet. Fiz. Rinkins, 13, 495 (1973)
"MC approximation for the outer shells of atoms in the $3p^n$ and $4p^n$ configurations"

V. Kvasnika, V. Laurinc; Chem. Phys. Lett., 18, 375 (1973)
"MC HF Theory with Non-Orthogonal Orbitals"

B. Levy; Chem. Phys. Lett., 18, 59 (1973)
"MC SCF Wavefunctions for CH_4, C_2H_4 and C_2H_6"

H. Schlosser; Chem. Phys. Lett., 23, 545 (1973)
"MC SCF Theory of Localized Orbitals: Choice of Localization Potential"

W.J. Stevens, F.P. Billingsley, Phys. Rev., A8, 2236 (1973)
"Coupled MC SCF Method for Atomic Dipole Polarizabilities. II. Application to First-Row Atoms, Li through Na"

M.H. Wood, A. Veillard; Mol. Phys., 26, 595 (1973)
"On Convergence Guarantees for the MC SCF Theory"

1974

G.B. Bacskay, N.S. Hush; Theoret. Chim. Acta, 32, 311 (1974)
"Molecules in Electric Fields. I. The Polarizability of the Hydrogen Molecules"

F.P. Billingsley, M. Krauss; J. Chem. Phys., 60, 4130 (1974)
"MC SCF Calculation of the dipole moment function of $Co(X^1\Sigma^+)$*"

T.Ch. Chang; Diss. Abstr. Int. B., 34, 5844 (1974)
"New methods for optimizing atomic and molecular MC wavefunctions"

T.K. Das, M.T. Coelho; Phys. Rev., C10, 1574 (1974)
"MC SCF method applied to an exactly solvable model"

G. Das, T. Janis, A.C. Wahl; J. Chem. Phys., 61, 1274 (1974)
"Ground and Excited States of the Diatoms CN and AlO"

C.W. Eaker, J. Hinze; J. Am. Chem. Soc., 96, 4084 (1974)
"Semiempirical MC-SCF Theory. I. Closed Shell Ground State Molecules"

C. Froese-Fischer; Int. J. Quant. Chem. Symp., 8, 5 (1974)
"A MC HF Approach to Atomic Structure Calculations"

Ch. Froese-Fischer, K.M.S. Saxena; Phys. Rev., A9, 1498 (1974)
"Correlation study of Be $1s^2 2s^2$ by a separated electron part numerical MC HF procedure"

T.L. Gilbert; J. Chem. Phys., 60, 3835 (1974)
"MC SCF Theory for Localized Orbitals. II. Overlap Constraints, Lagrangian Multipliers, and the Screened Interaction Field"

A. Golebiewski, E. Novak-Broclawik; Mol. Phys., 28, 1283 (1974)
"On the MC SCF Theory of Closed-Shell Systems. II. Two-by-Two Rotation Method and its applications"

F. Grein, A. Banerjee; Chem. Phys. Lett., 25, 255 (1974)
"Comparison of singly excited and pair-excited MC wavefunctions for the ground states of the atoms He to F"

M. Hackmeyer; Int. J. Quant. Chem., 8, 783 (1974)
"A general Ab Initio Molecular MC SCF Algorithm"

S. Iwata, K. Morokuma; Theoret. Chim. Acta, 33, 285 (1974)
"MC Electron-Hole Potential Method for Excited States"

O. Kikuchi, K. Aoki; Bull. Chem. Soc. Jap., 47, 2915 (1974)
"The Application of an MC-SCF Theory to Organic Chemical Reactions"

M. Krauss, D. Neumann; Mol. Phys., 27, 917 (1974)
"MC SCF Calculation of the Dissociation Energy and Electronic Structure of FH"

V.A. Kuprievich, O.V. Shramko, V.E. Klimenko; Theor. Eksp. Khim., 10, 746 (1974)
"Ground state of LiH molecule studied by the MC SCF theory"

G.C. Lie; J. Chem. Phys., 60, 2991 (1974)
"Study of the Theoretical Dipole Moment Function and IR transition Matrix for the x $^1\Sigma^+$ State of the HF Molecule"

I. Lindgren; J. Phys. (B) Atom. Molec. Phys., 7, 2441 (1974)
"The Rayleigh-Schrodinger Perturbation and the Linked-Diagram Theorem for a MC Model Space"

N.G. Mukherjee; Chem. Phys. Lett., 24, 441 (1974)
"Variational procedure for the optimization of MCSCF orbitals"

D. Mukherjee; Int. J. Quant. Chem., 8, 247 (1974)
"A Comparative Calculation on Excited State Energies of Some Conjugated Hydrocarbons"

M. Schlosser; J. Chem. Phys., 61, 2814 (1974)
"Simplified MC SCF Theory for Localized Orbitals. III. Evaluation of Operators and Matrix Elements"

W.J. Stevens, A.C. Wahl, M.A. Gardner, A.M. Karo; J. Chem. Phys., 60, 2195 (1974)
"Ab Initio Calculation of the Ne-Ne $^1\Sigma^+_g$ Potential at Intermediate Separations"

A.F. Wagner, G. Das, A.C. Wahl; J. Chem. Phys., 60, 1885 (1974) "Calculated long-range Interactions and Low Energy Scattering of Ar-H"

M.H. Wood; Chem. Phys. Lett., 24, 239 (1974) "The Barrier to Rotation for the Ground State of Ethylene: A DC SCF Approach"

1975

R. Ahlrichs, F. Driessler; Theoret. Chim. Acta, 36, 275 (1975) "Direct Determination of Pair Natural Orbitals: A New Method to Solve the MC HF Problem for Two-Electron Wave Functions"

A.V. Anataraman; Disst. Abstr. Int. B, 35, 4406 (1975) "MC SCF theory using non-orthogonal orbitals: Be atom"

F.P. Billingsley II; J. Chem. Phys., 62, 864 (1975) "MCSCF Calculation of the Dipole Moment Function and Potential Curve of $NO(x^2\Pi)$"

F.P. Billingsley II; J. Chem. Phys., 63, 2267 (1975) "Improved MCSCF Dipole Moment Function for $NO(x^2\Pi)$"

J.W. Birks, H.S. Johnston, H.F. Schaefer III; J. Chem. Phys., 63, 1741 (1975) "Ne-H-H Potential Energy Surface Including Electron Correlation"

D.B. Cook; Mol. Phys., 30, 733 (1975) "Doubly-Occupied Orbital MCSCF Methods. I. The PEMCSCF Method: Formulation and Application to Small Molecules"

P.D. Dacre, C.J. Watts, G.R.J. Williams, R. McWeeny; Mol. Phys., 30, 1203 (1975) "Molecular MCSCF Calculations by Direct Minimization. I. The Single Excitation MCSCF Method"

Ch. Froese-Fischer; Int. J. Quant. Chem., 9, 273 (1975) "A Correlation Study of Li Ground State by the MC HF Procedure"

Ch. Froese-Fischer, K.M.S. Saxena; Phys. Rev., A12, 2281 (1975) "Correlation study of the Be $1s^2 2s2p\ {}^1P$ by a separated-part numerical MC HF procedure"

G.D. Gillispie; Diss. Abst. Int. B, 36, 1285 (1975) "Electronic structure of NO_2. MC SCF calculation of the low-lying electronic states. Spectral interpretation"

G.D. Gillispie; A.U. Khan, A.C. Wahl, R. Hosteny, M. Krauss; J. Chem. Phys., 63, 3425 (1975) "Electron structure of NO_2. I. MC SCF calculation of the low-lying electronic states"

F. Grein, A. Banerjee; Chem. Phys. Lett., 31, 281 (1975) "Variational Wavefunctions for Low-Lying Excited States"

F. Grein, A. Banerjee; Int. J. Quant. Chem. Symp., 9, 147 (1975) "A MC Method for Excited States of Atoms and Molecules"

G.A. Henderson, G. Das, A.C. Wahl; J. Chem. Phys., 63, 2805 (1975) "MC Studies of Some Low-Lying Bound States of VH"

V.A. Kuprievich, O.V. Shramko; Int. J. Quant. Chem., 9, 1009 (1975) "Improved Convergence of SC Procedures in the MC SCF Theory"

E.A. McCullough Jr.; J. Chem. Phys., 62, 3991 (1975) "The Partial-Wave SCF Method for Diatomic Molecules: Computational Formalism and Results for Small Molecules"

M. Nakoyama, K. Sazi, Y.J. Ihaya; Theoret. Chim. Acta, 38, 327 (1975)
"Predictions of molecular geometries and electronic spectra of complex unsaturated molecules from MC LCAO MO"

S. Polezzo; Theoret. Chim. Acta, 40, 245 (1975)
"Direct minimization of the Energy Functional in LCAO-MO Density Matrix Formalism. II. MC-SCF Theory for Ground and Excited States"

E.S. Sachs, J. Hinze, N.H. Sabelli; J. Chem. Phys., 62, 3367 (1975)
"MCSCF Calculations for six States of NaH"

H. Schlosser; Chem. Phys. Lett., 35, 239 (1975)
"MC SCF theory of localized orbitals. Secular Problem"

B.T. Stucliffe; Theoret. Chim. Acta, 39, 93 (1975)
"Convergence properties of direct energy minimization with respect to linear coefficients in the Mc LCAO SCF approach"

M. Yaszunski, A.J. Sadlej; Theoret. Chim. Acta, 40, 157 (1975)
"Coupled MCSCF Perturbation Theory"

1976

J.O. Arnold, E.E. Whiting, L.F. Sharbaugh; J. Chem. Phys., 64, 3251 (1976)
"A Nearly Exact MCSCF + CI Calculation of the Dissociation Energy of OH"

A. Benerjee, F. Grein; Int. J. Quant. Chem., 10, 123 (1976)
"Convergence Behavior of Some MC Methods"

M.A. Besson, M. Suard; Int. J. Quant. Chem., 10, 151 (1976)
"Analysis of the Space Correlation Factors in a MCSCF Representation of LiH and Li_2 Molecules"

L.M. Cheung; Diss. Abstr. Int. B, 36, 3404 (1976)
"The MC SCF method for many electron systems and its application to the dissociation of ethylene"

P.D. Dacre, C.J. Watts; Mol. Phys., 31, 1149 (1976)
"Molecular MCSCF Calculations by Direct Minimization. II. Calculations on H_2O using the Single Excitation MCSCF Method"

P. Fantucci, S. Polezzo, M.P. Stabilini; Theoret. Chim. Acta, 41, 311 (1976)
"Direct Minimization of the Energy Functional in LCAO-MO Density Matrix Formalism. III. All Single-Excitation Wavefunctions in MCSCF"

R.L. Jaffe, K. Morokuma; J. Chem. Phys., 64, 4881 (1976)
"MCSCF Potential Energy Surface for Photodissociation of H_2CO"

J. Kendrick, I.H. Hillier; Chem. Phys. Lett., 41, 283 (1976)
"A computational method of performing SCF calculations using bonded functions"

V.A. Kuprievich, V.E. Klimenko; Teor. Eksp. Khim., 12, 169 (1976)
"Computer program for optimizing multiconfiguration wavefunctions"

W. Meyer; J. Chem. Phys., 64, 2901 (1976)
"Theory of SC Electron Pairs. An Iterative Method for Correlated Many-Electron Wavefunctions"

R.P. Saxon, B. Liu; J. Chem. Phys., 64, 3291 (1976)
"Ab Initio Calculations of the $^3\Sigma^+_g$ and $^3\Sigma^+_\mu$ States of Singly Excited Ar_2"

W.H.E. Schwarz, T.C. Chang; Int. J. Quant. Chem. Symp., 10, 91 (1976)
"MC Wave Functions for Highly Excited States by the Generalized Brillouin Theorem Method"

L.G. Yaffe, W.A. Goddard III; Phys. Rev., A13, 1682 (1976)
"Orbital Optimization in Electronic Wave Functions; Equations for Quadratic and Cubic Convergence of General MC Wave Functions"

1977

A. Banerjee, F. Grein; J. Chem. Phys., 66, 1054 (1977)
"MC wavefunctions for excited states. Selection of optimal configurations: the $b^1\Sigma^+$ and $d^1\Sigma^-$ states of NH"

R. Carbo, J.A. Hernandez; Chem. Phys. Lett., 47, 85 (1977)
"A General Multiconfigurational Paired Excitation SCF Theory (MC PE SCF)"

T.C. Chang, W.H.E. Schwarz; Theoret. Chim. Acta, 44, 45 (1977)
"Generalized Brillouin Theorem MC method for excited states"

K. Ishida, K. Kondo, T. Yonezawa; J. Chem. Phys., 66, 2883 (1977)
"Studies on bond dissociation in CH_4, NH_3, and H_2O by the MC SCF method"

Subject Index